# Multi-system Imaging Spectrum associated with Neurologic Diseases

# Multi-system Imaging Spectrum associated with Neurologic Diseases

Edited by

**Bo Gao**
Department of Radiology, The Affiliated Hospital of Guizhou Medical University, Guizhou, China

**Alexander Lerner**
Keck Medical Center of USC, Los Angeles, CA, USA

ELSEVIER

**ACADEMIC PRESS**
An imprint of Elsevier

Academic Press is an imprint of Elsevier
125 London Wall, London EC2Y 5AS, United Kingdom
525 B Street, Suite 1650, San Diego, CA 92101, United States
50 Hampshire Street, 5th Floor, Cambridge, MA 02139, United States
The Boulevard, Langford Lane, Kidlington, Oxford OX5 1GB, United Kingdom

ISBN: 978-0-323-91795-7

For information on all Academic Press publications visit our website at
https://www.elsevier.com/books-and-journals

*Publisher:* Stacy Masucci
*Acquisitions Editor:* Elizabeth A.Brown
*Editorial Project Manager:* Michaela Realiza
*Production Project Manager:* Selvaraj Raviraj
*Cover Designer:* Vicky Pearson Esser

Typeset by TNQ Technologies

Working together
to grow libraries in
developing countries

www.elsevier.com • www.bookaid.org

# Contents

# Contributors

**Bihong T. Chen**, Department of Diagnostic Radiology, City of Hope National Medical Center, Duarte, CA, United States

**Gordon Crews**, Keck School of Medicine of USC, Los Angeles, CA, United States

**Xia Du**, Department of Radiology, The Affiliated Hospital of Guizhou Medical University, Guiyang, China

**Bo Gao**, Department of Radiology, The Affiliated Hospital of Guizhou Medical University, Guiyang, Guizhou, China

**Raymond Huang**, Keck School of Medicine of USC, Los Angeles, CA, United States

**Bin Huang**, Department of Radiology, The Affiliated Hospital of Guizhou Medical University, Guiyang, Guizhou, China

**Teng Jin**, Department of Radiology, Union Hospital, Huazhong University of Science and Technology, Wuhan, China

**Shi-ji Kan**, Department of Radiology, The Affiliated Hospital of Guizhou Medical University, Guiyang, Guizhou, China

**Pinggui Lei**, Department of Radiology, The Affiliated Hospital of Guizhou Medical University, Guiyang, Guizhou, China

**Alexander Lerner**, Keck School of Medicine of USC, Los Angeles, CA, United States

**Heng Liu**, Department of Radiology, The Affiliated Hospital of Zunyi Medical University, Zunyi, China

**Peng Liu**, Department of Radiology, Hunan Provincial People's Hospital, First Affiliated Hospital of Hunan Normal University, Changsha, Hunan, China

**Xiaoqing Liu**, Department of Radiology, The Affiliated Hospital of Zunyi Medical University, Zunyi, China

**Daniel Phung**, Keck School of Medicine of USC, Los Angeles, CA, United States

**Nasim Sheikh-Bahaei**, Keck School of Medicine of USC, Los Angeles, CA, United States

**Jing Wang**, Department of Radiology, Union Hospital, Huazhong University of Science and Technology, Wuhan, China

**Tian-le Wang**, Affiliated Hospital 2 of Nantong University, Nantong, Jiangsu, China

**Huiying Wu**, Department of Radiology, Guangzhou Women and Children's Medical Center, Guangzhou, China

**Xuntao Yin**, Department of Radiology, Guangzhou Women and Children's Medical Center, Guangzhou, China

**Jing Yu**, Department of Radiology, The Affiliated Hospital of Guizhou Medical University, GuiYang, China

# Biographies

Bo Gao, MD, PhD, Professor and Chair, Department of Radiology, The Affiliated Hospital of Guizhou Medical University, China

Dr. Gao is a Professor and Chair in the Department of Radiology at the Affiliated Hospital of Guizhou Medical University, in China. His three major areas of research interest are (1) radiomics, radiogenomics, and AI in cancer imaging; (2) functional MRI, diffusion tensor imaging, and permeability imaging in tumor theranostics; and (3) posterior reversible encephalopathy syndrome and associated entities. Dr. Gao has edited *Classic Imaging Signs* (2021) and *Imaging of CNS Infections and Neuroimmunology* (2019) both with Springer and has published over two dozen journal articles.

Alexander Lerner, MD, Associate Professor of Clinical Radiology, Program Director of USC/LAC+USC Neuroradiology Fellowship, Director of Clinical Functional MRI, Department of Radiology, Keck School of Medicine of University of Southern California, USA

Dr. Lerner is an Associate Professor of Clinical Radiology, Program Director of USC/LAC+USC Neuroradiology Fellowship and Director of Clinical Functional MRI in the Department of Radiology at Keck School of Medicine of the University of Southern California in USA. Dr. Lerner's research interests include advanced MR imaging using functional MRI, DTI, MR perfusion, and permeability imaging of the brain and spinal cord. He specializes in the diagnosis of disorders causing headaches, neurovascular compression disorders, CSF flow disorders, epilepsy, inflammatory, and degenerative disorders of the CNS. Dr. Lerner has been an invited speaker at multiple radiology societies at regional and national levels. He has also authored and coauthored multiple publications and book chapters.

# Foreword I

Systemic diseases with origins ranging from congenital, metabolic abnormalities to inflammation may affect the brain and head and neck, respiratory, digestive, excretory, or musculoskeletal system. Specifically, a variety of disorders, including infectious, inflammatory, hereditary, and metabolic diseases, may affect both the brain and abdominal cavity, and the findings in one region may help establish the diagnosis or limit the range of different diagnosis. Establishing an accurate early diagnosis enables clinicians to adequately manage these unusual diseases and potentially avert life-threatening complications.

This book written by Dr. Gao and his colleagues provides a multimodality review of common and uncommon multisystem clinical entities with radiography, CT, MRI, angiography, and other modalities. It is designed to enhance recognition of specific imaging, enabling the interpreter to confidently reach an accurate diagnosis. The materials included in this book were collected from renowned university hospitals and are well organized. This book aims to serve as a valuable review for residents, fellows, and trainees preparing for boards licensing test, and a trusted daily reference for practicing radiologists. Accompanying text explains the history and meaning of the descriptive or metaphoric sign. Uniquely written from a practical point of view, each case leads you through a radiology expert's thought process in analyzing imaging pitfalls of different organs or systems. These cases highlight clinical presentation, relevant pathology, anatomy, physiology, and pertinent imaging features of common disease processes. I wish this book would be an irreplaceable reference for reader confronted with the challenges of multisystem imaging interpretation.

Dr. Gao is a Professor and Chairman of the Department of Radiology, the Affiliated Hospital of Guizhou Medical University. His research interests mainly focus on diagnostic imaging especially neuroradiology. Over the past decades, Dr. Gao has participated in a couple of scientific research projects and has published over 60 peer-reviewed papers. He is the executive editor-in-chief of our new launched journal *iRADIOLOGY* and also the editor of more than 10 books. I am very glad to introduce his new works to the interested readers and hope to enrich their reading choice and enhance professional vision. I am also looking forward to the revised edition of this book in the near future.

**Professor Peng Luo**
*President of Guizhou Medical University*
September 1, 2022

# Foreword II

I have known and worked with Drs. Gao and Lerner for more than 13 years and it gives me great pleasure to write the foreword for their book *Multi-System Imaging Spectrum Associated with Neurologic Diseases*. Drs. Gao and Lerner are what you would call the academic triple threat: expert neuroimaging diagnosticians, researchers, and teachers. They are the people I would consult for difficult, troubling cases. They have assembled an international team of expert authors who will be providing a detailed but concise overview of the pathophysiology, clinical, and imaging findings throughout the various body systems that have neurologic manifestations.

There is a nice introduction into the reasoning and step-wise approach to arriving at the correct imaging diagnosis using both conventional and physiologic imaging. Each of the disease entities has a concise, but comprehensive description of both neural as well as extraneurological manifestations with nice imaging examples. There are 12 chapters which cover the nervous, muscular, respiratory, cardiovascular, digestive, excretory, reproductive, skeletal, endocrine, immune, lymphatic, and integumentary systems. These include relatively common entities, but there is coverage of rarer diseases including reviews of poliomyelitis and extradural meningiomas.

This is a very unique addition to the body of knowledge in neuroradiology. This will be valuable not only to the academic neuroradiologist subspecialist but to medical students, residents, fellows, and general radiologists. I believe it will be an enjoyable read and something that will be a value contribution to our field.

**Mark S. Shiroishi, MD, MS, FASFNR**
*Department of Radiology*
*Department of Population and Public Health Sciences*
*Keck School of Medicine*
*University of Southern California*
*Los Angeles, CA*
November 1, 2022

# Preface

The diversity of neurologic disorders associated with diseases involving the different organ systems is immense. However, understanding of this subject is critical for effective diagnosis and management of these complex cases. The intent in writing of this book is to provide a guide for diagnosis and evaluation for these various neurologic conditions in a systematic fashion. The text is organized by different organ systems and specific diseases arising in these regions. Neurologic disorders associated with these diseases are systematically reviewed and discussed. Appropriate imaging cases are provided to illustrate the typical pathologic manifestations encountered in these conditions.

This book includes a great number of disorders which may involve the nervous system and other various organ systems. While some disorders primarily involve the CNS, others have extensive systemic involvement with variable involvement of the nervous system. We have decided to focus on some of the more common and unique forms of these disorders where imaging plays a key role in evaluating the involvement of the nervous system.

Due to complexity of this subject, a truly comprehensive list of all the entities associated with multisystem involvement would be excessive and redundant. The goal of this book is to provide the reader with typical examples of the entities associated with diseases in each organ system and to include some important lesser-known conditions. This format should help the reader develop a framework for understanding and recognizing neurologic disorders associated with diseases involving or arising in other organ systems.

We hope that you find this book useful as a reference and as a guide in understanding the imaging and clinical manifestations of these complex neurologic conditions.

**Alexander Lerner MD**
*Associate Professor of Clinical Radiology*
*University of Southern California*
*Keck School of Medicine*
*Los Angeles, CA*

# Chapter 1

# Introduction

**Bo Gao[1], Shi-ji Kan[1] and Bihong T. Chen[2]**
[1]Department of Radiology, The Affiliated Hospital of Guizhou Medical University, Guiyang, Guizhou, China; [2]Department of Diagnostic Radiology, City of Hope National Medical Center, Duarte, CA, United States

## 1. Pathogenesis of multisystem diseases

The pathogenesis of multisystem diseases can stem from various causes. Congenital multisystem diseases may be associated with neurological diseases. These are largely due to chromosomal or genetic abnormalities. Due to the large number of genes on the chromosome and the pleiotropy of genes, chromosomal diseases often involve morphological and functional abnormalities of multiple organs and systems, and their clinical manifestations often present as syndromes. For example, von Hippel-Lindau (VHL) disease is a hereditary autosomal dominant neoplastic disease that is associated with various tumor types, including clear-cell renal cell carcinoma, central nervous system (CNS) and retinal hemangioblastomas, pheochromocytomas, pancreatic neuroendocrine tumors, and pancreatic and renal cysts. The VHL gene is located on 3p25.5 encoding a tumor suppressor, pVHL [1,2].

For inflammatory diseases, their mechanisms involve exogenous and endogenous damage factors entering the nervous system via other tissues. There are local and systemic responses associated with this process which may produce a series of reactions in an attempt to eliminate the offending agents, absorb the necrotic tissue, and repair the damage. Consequently, this complex process may cause lesions in the nervous system and other organs. There are many types of inflammatory factors, with most common ones listed as follows:

(1) Physical factors such as extreme high or low temperature, mechanical trauma, ultraviolet rays, and ionizing radiation.
(2) Chemical factors including both exogenous and endogenous compounds. Exogenous chemical compounds include strong acid, base, oxidant, mustard gas, etc. Endogenous chemicals include decomposition products of internal tissues and metabolites accumulated in the body such as urea due to pathological conditions.
(3) Biological factors are most common, and include agents such as bacteria, fungi, viruses, mycoplasma, protozoa, parasites, etc.
(4) Tissue necrosis can be caused by ischemia or hypoxia. Necrotic tissue is also a potential inflammatory factor.
(5) Allergic disorders are caused by an abnormal immune state leading to improper immune response to benign environmental or internal factors and resulting in tissue damage and inflammation. One example is neuromyelitis optica (NMO; also known as Devic's syndrome), which is an inflammatory CNS syndrome characterized by attacks of acute optic neuritis and transverse myelitis. In most patients, NMO is caused by pathogenetic serum IgG autoantibodies to aquaporin 4, the most abundant water-channel protein in CNS. All factors that can cause tissue and cell damage may give rise to inflammation [3,4].

Metabolic and endocrine diseases can be related in etiology. Metabolism is the general term for the synthesis and decomposition of substances in the body, which is the cornerstone of human physiology. From birth, the body needs to absorb nutrients and expel metabolic waste in order to keep the individual alive and to allow them to thrive and reproduce. The human body is always exchanging material with the outside world. Metabolic disorders occur when a sequence of metabolic processes is impaired or disrupted, resulting in deficit or excessive accumulation of substances that can result in damage to local tissues, organs, or systems throughout body. One example of this category is multiple endocrine neoplasia (MEN) syndromes which includes MEN1, MEN2 (formerly MEN2A), MEN3 (formerly MEN2B), and recently identified MEN4. MEN syndromes are autosomal dominant and are distinguished phenotypically by the development of synchronous or metachronous tumors in endocrine glands [5]. Clinical presentations are varied and often related to overproduction of specific hormones. Common tumors in MEN include parathyroid adenoma, pancreatic NETs, and pituitary adenoma. Excessive parathyroid hormone secretion can cause calcium and phosphorus metabolism disorders. Lesions in anterior pituitary gland (adenohypophysis) may lead to the abnormal secretion of hormones such as growth hormone, thyroid-stimulating hormone, and adrenocorticotropic hormone, resulting in disorders involving the nervous system and other organs.

In this book, we discussed the characteristic imaging appearances of neurological disorders manifested in multisystem diseases involving the head and neck, thoracic, abdominal cavity, pelvis, or musculoskeletal system. A brief overview of the clinical features of each disease is discussed with its neurologic findings in the context of multisystem involvement.

# 2. Diagnostic algorithm of multisystem diseases

Hippocrates said a medical scientist must also be a philosopher, as clinical thinking can help to determine the correct diagnosis of disease, while erroneous or inappropriate clinical thinking would lead to delay in diagnosis and treatment.

Getting the right diagnosis is a key aspect of health care, which may provide an explanation of a patient's health problem and inform subsequent health care decisions. The diagnostic process is a complex, collaborative activity that involves clinical reasoning and information gathering. Diagnostic errors with inaccurate or delayed diagnoses persist throughout all settings of care and continue to harm patients. Diagnostic errors contribute to as many as 70% of medical errors. Prevention of diagnostic errors is more complex than building safety checks into health care systems. It requires an understanding of critical thinking, clinical reasoning, and cognitive processes through which diagnosis is made. When diagnostic error is recognized, it is imperative to identify where and how mistakes in clinical reasoning occur. Cognitive biases may contribute to errors in clinical reasoning. Cognitive bias awareness training and debiasing strategies should be developed to decrease diagnostic errors [6,7]. In order to decrease the rate of errors, it is essential for clinicians to use critical thinking which plays an important role in imaging diagnosis. Critical thinking is a cognitive process used for arriving at the correct diagnosis. Clinical reasoning involves cognitive processes used for analyzing knowledge relevant to a clinical situation or specific patient [8].

## 2.1 Overall approach in imaging diagnosis

The appropriate clinical reasoning is critical thinking, which is the thinking process of continuous analysis, confirmation, supplementation, and modification of clinical decision-making. The clinician should perform a comprehensive review of the disease based on the patient's medical history and physical examination, and then use imaging to further investigate the disease. The thought process of physicians for diagnosis and differentiation of diseases is usually as follows. First, lesions are identified, and their imaging characteristics are noted. Imaging features can quickly define the type and scope of the disease investigated and help make the correct diagnosis or narrow differential diagnosis based on the clinical data. We refer to this as a triad, "site of disease, imaging appearance, and clinical manifestations" (or "new triad"). Finally, one can identify and judge the anatomic, pathologic, and physiologic changes associated with these findings and put forward a preliminary diagnosis and differential diagnosis. Scientific clinical thinking should be a comprehensive thought process integrating clinical and imaging information (Fig. 1.1). When considering how to prevent diagnostic errors that are due to cognitive processes, it is important to understand how physicians make clinical decisions [6].

Typically, clinicians and radiologists may be familiar with their own professional reasoning process, while each does not recognize elements of reasoning by the other side. Imaging changes and clinical manifestations reflect different aspects of the disease. Only a comprehensive analysis of both aspects can help to arrive at a correct diagnosis. If one only pays attention to the imaging aspects and ignore clinical aspects, excluding important medical history, they may make a wrong diagnosis based on imaging. Imaging also has significant limitations. Pathological changes manifested on imaging sometimes may not be consistent with the clinical course, which may be referred to as "the separation of clinical manifestation and imaging appearance." Each imaging examination is limited to specific anatomical site and has specific tissue contrast limitations, which may lead to missed diagnosis.

## 2.2 Thinking process in imaging diagnosis

Imaging diagnosis is based on the image characteristics of the disease. This process uses the knowledge of anatomy, pathology, and etiology of disease in combination with the clinical manifestations to comprehensively analyze the nature and scope of the disease and to arrive at the most specific diagnosis possible. Imaging diagnostic thinking also has a specific paradigm and characteristics. Generally, the reasoning associated with imaging diagnosis is guided by the spatial

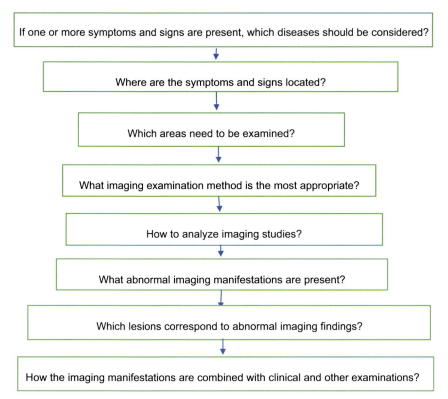

If one or more symptoms and signs are present, which diseases should be considered?

Where are the symptoms and signs located?

Which areas need to be examined?

What imaging examination method is the most appropriate?

How to analyze imaging studies?

What abnormal imaging manifestations are present?

Which lesions correspond to abnormal imaging findings?

How the imaging manifestations are combined with clinical and other examinations?

**FIGURE 1.1**  Representative steps for clinical imaging comprehensive analysis.

reasoning and combined with logical reasoning of temporal course. At the same time, empirical thinking and dialectical thinking are involved (Fig. 1.2). The core principal is to assure the safety of patient. Only by the optimal cross-application of various forms of thinking can the best diagnostic conclusions be identified. Optimal images are the basis of imaging—diagnostic correlation. Accurate clinical history and relevant laboratory tests are also critical factors in the diagnostic process. Radiologists must broaden their thinking when facing with diseases that may have identical or similar imaging presentations.

For example, rim-enhancing lesions in the brain can be seen in brain metastases, high-grade glioma, brain abscess, brain tuberculosis, and other diseases. It is challenging to generate a differential diagnosis only based on this imaging characteristics. Accurate diagnosis is unlikely without analysis of clinical manifestations and laboratory examination. Comprehensive analysis is crucial to achieve early diagnosis and timely treatment. One study shows that abscess was significantly different from glioblastoma and metastasis on quantitative and qualitative diffusion tensor imaging analysis of the cystic lesions, enhancing rims, and perilesional edema, respectively. These differences indicate how different patterns of water diffusion may allow discrimination of abscess from glioblastoma and metastasis. Therefore, this advanced imaging approach may improve specificity of imaging diagnosis [9].

Diffusion-weighted imaging (DWI) exploits random motion of water molecules that is more sensitive in detection of early cellular or intracellular changes than T2-weighted signal intensity. There are many lesions with elevated intensities on DWI, such as acute infarction [10]. Restricted diffusion is also typical of epidermoid cysts and may even be seen in some demyelinating lesions [11,12]. Radiologists should consider this information and then differentiate these lesions according to clinical and/or imaging information that acute infarction may show diffusion restriction in the center of lesions unlike demyelinating diseases in which diffusion restriction is mainly at the periphery of the lesion.

Multisystem disease associated with neurological disorder is similar to a large information network. It is difficult to make accurate judgment without rigorous logical reasoning and understanding of clinical thinking.

## 3. The role of imaging in diagnosing multisystem diseases

Imaging has made dramatic progress over the last decades. In the recent decades, rapid development of imaging technologies (such as CT, MRI, DSA, SPECT, PET, PET-CT, PET-MRI, etc.) has allowed for accurate early diagnosis of

**FIGURE 1.2**   Representative clinical imaging diagnostic algorithm.

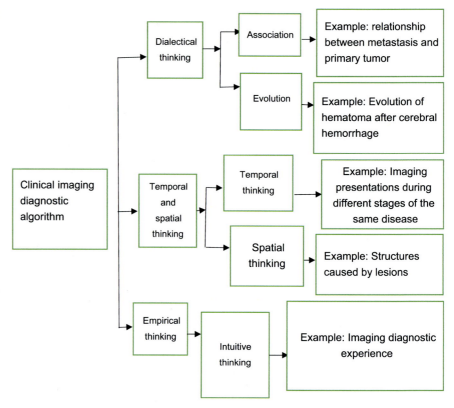

various neurological diseases. Imaging has also become important for evaluation of drug efficacy and prognosis. Here, we use tuberous sclerosis as an example to illustrate the role of imaging in clinical practice.

## 3.1 Overview

Tuberous sclerosis complex (TSC) is a neurocutaneous syndrome that can present at any age and can affect multiple organ systems (Table 1.1). This disorder is usually identified in infants and children based on characteristic skin lesions, seizures, and cellular overgrowth or hamartomas in the heart, brain, and kidneys [13].

**TABLE 1.1** Clinical and imaging features of tuberous sclerosis.

| Impact system | Major disease | Imaging findings |
| --- | --- | --- |
| Central nervous system<br>Respiratory system | Subependymal nodules, cortical and subcortical tubers, cerebral white matter radial migration lines, and subependymal giant cell astrocytoma<br>Angiomyolipoma<br>Lymphangioleiomyomatosis | Calcified subependymal nodules, multiple hyperintense cortical and subcortical tubers, enhancing subependymal giant cell astrocytoma<br>Variable density<br>Multiple thin-walled cysts surrounded by normal lung parenchyma |
| Cardiovascular system<br>Urinary system | Cardiac rhabdomyomas<br>Angiomyolipoma | Multiple echogenic intracardiac masses<br>Rounded lesion with fat |
| Musculoskeletal system | Focal sclerotic lesions, bone cysts, and periosteal new bone formation | Increased or reduced intensity |

## 3.2 Pathophysiology

Tuberous sclerosis is an autosomal dominant disorder, although two-thirds of patients have sporadic mutations. The genes in which abnormalities are called TSC1 and TSC2 [14].

## 3.3 Central Nervous System involvement

Imaging studies can show cortical and subcortical tubers and migration lines in white matter, these being the manifestations responsible for cortical dysplasia. Other findings include subependymal nodules in the wall of lateral ventricles (Fig. 1.3) and third ventricle. In 80% of the cases, the patients are asymptomatic. Giant cell subependymal astrocytomas are observed in 5%—15% of TSC patients, which in turn lead to ventriculomegaly and hydrocephalus, resulting in marked morbidity and mortality [15].

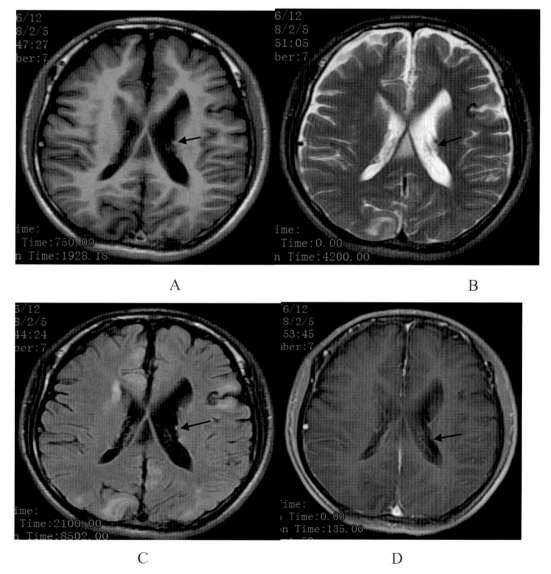

FIGURE 1.3  A 24-year-old man has tuberous sclerosis with subependymal nodules in the left lateral ventricle (arrows). Figures A—D show the subependymal nodules in T1-weighted imaging (A), T2-weighted imaging (B), fluid-attenuated inversion recovery (FLAIR) imaging (C), and contrast-enhanced imaging (D).

## 3.4 Urinary system

Renal complications are the second most common cause of morbidity and mortality in TSC patients. Angiomyolipomas are benign hamartomatous neoplasms variably composed of smooth muscle, fat, and abnormal blood vessels (Fig. 1.4) [16].

## 3.5 Digestive system

Hepatic angiomyolipomas are the most common hepatic manifestation of tuberous sclerosis. An estimated 16%—24% of patients with tuberous sclerosis have hepatic angiomyolipomas. Only 5.8% of hepatic angiomyolipomas have been found to be associated with tuberous sclerosis. Most of these lesions are asymptomatic; however, patients may present with abdominal pain or abdominal mass. Like renal angiomyolipomas, hepatic angiomyolipomas consist of intratumoral macroscopic fat, which appears as fat attenuation on CT images (Fig. 1.5) [17].

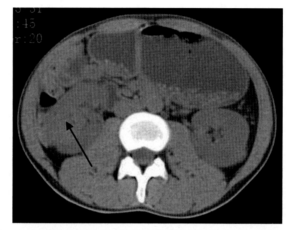

**FIGURE 1.4** CT image showing a right renal angiomyolipoma containing fat and smooth muscle (arrow).

**FIGURE 1.5** A 24-year-old man has tuberous sclerosis with a hepatic angiomyolipoma containing fat (arrow).

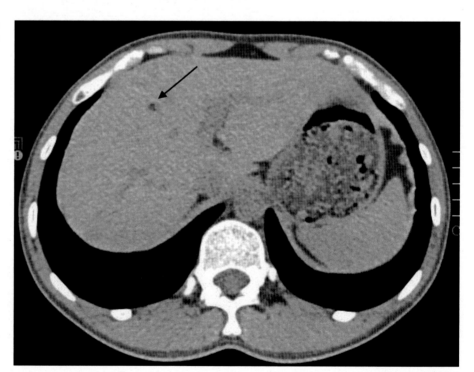

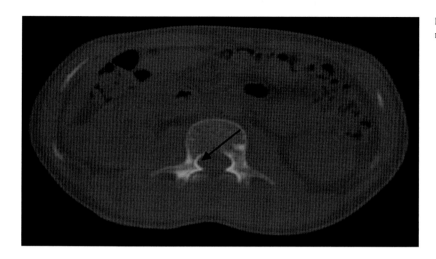

**FIGURE 1.6** A 24-year-old man has tuberous sclerosis and a sclerotic bone lesion in a vertebra (arrow).

## 3.6 Musculoskeletal system

Osseous manifestations are variable and consist of focal sclerotic lesions (Fig. 1.6), bone cysts, and periosteal new bone formation. Focal sclerotic lesions are found in the vertebrae, ribs, and the iliac aspect of the sacroiliac joints. Bone cysts are most commonly visualized in the phalanges of the hands and feet. The significance of these findings is unknown. As such, there are currently no recommendations for follow-up or intervention.

## 3.7 Respiratory system

The pulmonary manifestation of tuberous sclerosis is lymphangioleiomyomatosis, a proliferation of smooth muscle cells in the lymphatics accompanied by cystic changes in the lung parenchyma. Lymphangioleiomyomatosis can be seen as multiple thin-walled cysts scattered diffusely and surrounded by normal lung parenchyma.

The clinical diagnostic criteria of tuberous sclerosis are divided into major and minor criteria. Major criteria include the following: hypopigmented macules ($\geq$3, with at least 5 mm diameter), angiofibromas ($\geq$3) or fibrous cephalic plaque, ungual fibromas ($\geq$2), shagreen patch, multiple retinal hamartomas, cortical dysplasia, subependymal nodules, subependymal giant cell astrocytoma, cardiac rhabdomyoma, lymphangioleiomyomatosis, and angiomyolipomas ($\geq$2). Minor criteria include the following: "confetti" lesion, enamel pits (>3), intraoral fibroma ($\geq$2), retinal hypopigmented macule, multiple renal cysts, and nonrenal hamartomas. The definitive diagnosis is defined by the presence of two major criteria or one major and two minor criteria. Probable diagnosis is made with one major criterion or two or more minor criteria (Table 1.2).

**TABLE 1.2 Clinical criteria for the diagnosis of tuberous sclerosis complex.**

| Major criteria | Minor criteria |
|---|---|
| 1. Hypomelanotic macules ($\geq$3, $\geq$ 5 mm in diameter) | 1. "Confetti" skin lesions |
| 2. Angiofibromas ($\geq$3) or fibrous cephalic plaque | 2. Dental enamel pits (>3) |
| 3. Ungual fibromas ($\geq$2) | 3. Intraoral fibroma ($\geq$2) |
| 4. Shagreen patch | 4. Retinal achromic patch |
| 5. Multiple retinal hamartomas | 5. Multiple renal cysts |
| 6. Cortical dysplasia | 6. Nonrenal hamartomas |
| 7. Subependymal nodules | |
| 8. Subependymal giant cell astrocytoma | |
| 9. Cardiac rhabdomyoma | |
| 10. Lymphangioleiomyomatosis | |
| 11. Angiomyolipomas ($\geq$2) | |

The clinical criteria for the diagnosis of TSC when combined with the above-described multisystem imaging appearances and other manifestations can help with the correct diagnosis of this disease. In addition, genetic testing can also be used to confirm the diagnosis.

A multimodality-based approach to imaging is essential in clinical practice. Advances in imaging have propelled imaging into delving into the biological, functional, and hemodynamic aspects of multiple pathophysiologic processes [18]. Current and future trends in medical imaging will focus on improving sensitivity and specificity in early diagnosis of various diseases, and to improve its use in clinical decision-making. The role of the radiologist is to combine the information from anatomic and functional imaging and assess responses to treatment [19]. Imaging is also increasingly used to quantify the effect of novel therapies.

# References

[1] Gossage L, Eisen T, Maher ER. VHL, the story of a tumour suppressor gene. Nat Rev Cancer 2015;15(1):55—64.

[2] Peng S, et al. The VHL/HIF Axis in the development and treatment of pheochromocytoma/paraganglioma. Front Endocrinol 2020;11:586857.

[3] Jarius S, et al. Neuromyelitis optica. Nat Rev Dis Prim 2020;6(1):85.

[4] Wu Y, Zhong L, Geng J. Neuromyelitis optica spectrum disorder: pathogenesis, treatment, and experimental models. Mult Scler Relat Disord 2019;27:412—8.

[5] McDonnell JE, et al. Multiple endocrine neoplasia: an update. Intern Med J 2019;49(8):954—61.

[6] Royce CS, Hayes MM, Schwartzstein RM. Teaching critical thinking: a case for instruction in cognitive biases to reduce diagnostic errors and improve patient safety. Acad Med 2019;94(2):187—94.

[7] C., et al. Committee on Diagnostic Error in Health. In: Balogh EP, Miller BT, Ball JR, editors. Improving diagnosis in health care. Washington (DC): National Academies Press (US) Copyright; 2015. 2015 by the National Academy of Sciences. All rights reserved.:

[8] Victor-Chmil J. Critical thinking versus clinical reasoning versus clinical judgment: differential diagnosis. Nurse Educat 2013;38(1):34—6.

[9] Toh CH, et al. Differentiation of brain abscesses from necrotic glioblastomas and cystic metastatic brain tumors with diffusion tensor imaging. AJNR Am J Neuroradiol 2011;32(9):1646—51.

[10] Nakajo Y, et al. Early detection of cerebral infarction after focal ischemia using a new MRI indicator. Mol Neurobiol 2019;56(1):658—70.

[11] Razek A, Elsebaie NA. Imaging of fulminant demyelinating disorders of the central nervous system. J Comput Assist Tomogr 2020;44(2):248—54.

[12] Hoang VT, et al. Overview of epidermoid cyst. Eur J Radiol Open 2019;6:291—301.

[13] Randle SC. Tuberous sclerosis complex: a review. Pediatr Ann 2017;46(4):e166—71.

[14] Curatolo P, Bombardieri R, Jozwiak S. Tuberous sclerosis. Lancet 2008;372(9639):657—68.

[15] Krishnan A, Kaza RK, Vummidi DR. Cross-sectional imaging review of tuberous sclerosis. Radiol Clin 2016;54(3):423—40.

[16] Baskin Jr HJ. The pathogenesis and imaging of the tuberous sclerosis complex. Pediatr Radiol 2008;38(9):936—52.

[17] Manoukian SB, Kowal DJ. Comprehensive imaging manifestations of tuberous sclerosis. AJR Am J Roentgenol 2015;204(5):933—43.

[18] Fernández-Friera L, García-Álvarez A, Ibáñez B. Imagining the future of diagnostic imaging. Rev Esp Cardiol 2013;66(2):134—43.

[19] Jayaprakasam VS, et al. Role of imaging in esophageal cancer management in 2020: update for radiologists. AJR Am J Roentgenol 2020;215(5):1072—84.

Chapter 2

# Nervous system

**Heng Liu and Xiaoqing Liu**
*Department of Radiology, The Affiliated Hospital of Zunyi Medical University, Zunyi, China*

Chapter 2.1

# Multisystem effects of Sturge—Weber Syndrome

## 1. Introduction

The somatic mutation of GNAQ gene leads to Sturge—Weber syndrome (SWS), which is a rare sporadic neurocutaneous syndrome. The incidence of SWS is the third in all neurocutaneous syndromes, and the first two are neurofibromatosis (NF) and tuberous sclerosis complex (TSC), respectively. The disease is characterized by facial port-wine stain (PWS) combined with choroidal hemangioma and cerebral dura or pia mater hemangioma [1—3]. Facial port-wine birthmarks can be seen in 90% of patients; among them 30%—70% have choroidal hemangioma [2]. The incidence of SWS is one case per 20,000—50,000 live births, and there is no gender and racial preference [1]. Clinical manifestations in the central system include epilepsy, migraine, fluctuating hemiplegia, developmental retardation, and stroke-like seizures. Additionally other systemic and organ lesions can also occur at the same time (Table 2.1.1). Many clinicians often fail to realize the disease as it is rare and complex [4]. In order to improve physician's understanding of the disease and to provide better diagnosis and treatment services for patients, we will introduce the multisystem complications of SWS here (Fig. 2.1.1).

## 2. Pathogenesis

It is not entirely clear why these lesions occur in SWS. At present, there are two main hypotheses about the pathogenesis of SWS/PWS, namely neural denervation and gene mutation. So far, we have known that mutations in GNAQ and GNB2 genes can lead to SWS [5,6]. SWS is a neurovascular nevus hemangioma disease caused by abnormal development of vascular nerve in the embryonic stage. During the first 3 months of fetal development, primitive blood vessels invade the

**TABLE 2.1.1 The accumulation of Sturge—Weber Syndrome in each system.**

| Impact system | Major disease | Image findings |
| --- | --- | --- |
| Central nervous system | Epilepsy, headache, stroke-like episodes, brain atrophy | Meningeal hemangioma, cortex/subcortex calcification |
| Skin | Port-wine stain | |
| Eyes | Glaucoma, choroidal hemangioma | Eyeball enlargement, choroidal hemangioma |
| Endocrine system | Growth hormone deficiency | Central hypothyroidism |

Multi-system Imaging Spectrum associated with Neurologic Diseases. https://doi.org/10.1016/B978-0-323-91795-7.00004-X

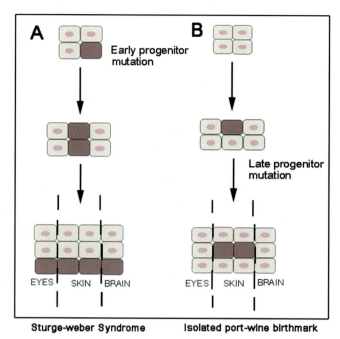

**FIGURE 2.1.1** (A) When somatic mutations occur in early progenitor cells, this mutation will affect multiple structures including the brain, skin, and eyes. (B) When the somatic mutation of GNAQ occurs in late progenitor cells, the effect of mutation is limited, which is manifested as isolated wine birthmarks, as shown in the figure. Depending on the location of subsequent somatic mutations, the affected structure may be just the brain or just the eyes. *The picture was modified from Comi AM. Sturge–Weber syndrome. Handb Clin Neurol 2015;132:157–68.*

developing brain, skin, and eyes, and somatic mutations may disrupt the normal vascular maturation [7]. It is speculated that somatic mutations lead to an abnormal development of the original embryonic venous plexus at 5–8 weeks of pregnancy, resulting in developmental vascular malformation, venous hypertension, and tissue hypertrophy. These malformed blood vessels have the same mitotic activity as the normal endothelial cells and grow with the body and cannot degenerate, resulting in increased vascular permeability, blood stasis, thrombosis and local ischemia, and ultimately a variety of systemic symptoms [8].

Cortical calcification is another pathological change characteristic of SWS, and lesions only occur in adjacent areas of hemangioma. We can observe that in SWS the brain tissue shows cortical calcium deposition, vascular dysplasia, glial hyperplasia, and sometimes neuronal loss or focal cortical dysplasia. Hypoxia causes damage to endothelial cells or epithelial cells, which results in calcification. Calcification may occur due to increased vascular permeability, as both pia mater enhancement and deep vein dilation indicate disruption of the blood-brain barrier. Previous studies have pointed out that PWS vessels lack peripheral blood vessels. Compared with the normal development of blood vessels, in SWS the nerve that controls the nerve cortical blood vessels only has norepinephrine sympathetic nerve fibers, and the expression of endothelin-1 in SWS deformed blood vessels is increased. These studies have shown that vasoconstrictive tension may increase in SWS cerebrovascular malformation. However, it is unclear whether these changes are pathological or compensatory [7].

## 3. Central nervous system

SWS is related to central nervous system (CNS) abnormalities, including epilepsy, stroke-like episodes, headache, and developmental retardation. Epilepsy is the most common neurological symptom in children with SWS. The seizure rate varies from 70% to 90% due to the degree of brain involvement. The incidence of epilepsy increases with age. 75% of seizures occur within 1 year old, 85% within 2 years old, and 90%–95% at 5 years old [9,10].

Imaging is essential in the diagnosis, detection, and follow-up of patients with SWS, of which MRI is currently considered to be the most superior examination method [11]. The best imaging methods to diagnose intracranial lesions in

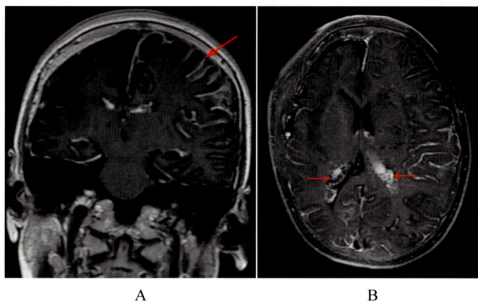

A                                               B

**FIGURE 2.1.2**   (A) Postcontrast coronal T1WI shows leptomeningeal enhancement on the left side (*arrow*); (B) Postcontrast axial T1WI reveals bilateral enlargement of the ipsilateral choroid plexus (*arrow*).

patients with SWS are T1-weighted gadolinium contrast MRI and magnetization-weighted imaging, both of which are particularly important in asymptomatic cases and are considered to be the standard evaluation for diagnosing intracranial lesions in SWS. MRI shows enhancement of leptomeningeal hemangioma, enlargement of medullary vein and periventricular veins, expansion and enlargement of choroid plexus on the affected side, as well as dilatation of the deep venous drainage vessels. Cortical atrophy and cortical calcifications beneath leptomeningeal hemangiomas are also typical MRI manifestations [8,9]. A direct feature of SWS patients is meningeal hemangioma [12], which exists in 98% of patients. The occipital and parietal pia matter, which are ipsilateral to PWS, are favored sites for meningeal hemangioma [8]. The presence and extent of pia mater vascular malformation, choroid plexus enlargement, deep medullary vein, and ependymal vein dilatation are often evaluated by postcontrast MRI [11]. On postcontrast T1WI, the leptomeningeal vessels are abnormally increased and dilated, showing obvious enhancement in the cerebral gyri, located on the surface of the cerebral gyri and cerebral sulci. Typically, intracranial angiomatosis is unilateral, while bilateral angiomatosis accounts for 15% [9]. At the same time, the children are often accompanied by abnormal enlargement of choroid plexus of the affected side, which show abnormally high signal on enhanced MRI (Fig. 2.1.2). It is worth noting that in newborns and young infants, MRI may appear to be negative, which may be due to venous stasis and hemodynamic damage that has not yet developed [7,11].

In the parenchyma, there is a range of atrophy with calcium deposition in the cortex and subcortex around the blood vessels [2]. The degree of meningeal enhancement is usually stable, while atrophy and calcification usually show intermittent progress [10,13]. CT scan can better display the calcifications of SWS than MRI, showing cortical or subcortical calcifications with "tram track" appearance or gyri-form high-density images on the surface of cerebral hemisphere (Fig. 2.1.3), and the lesions only occur in the adjacent area of the hemangioma [3,9]. On MRI, calcifications show low signal on T1WI and T2WI; however, it can be difficult to detect calcifications on T1-weighted images. On T2-weighted images, especially on T2*-weighted images, calcifications are easy to find, as T2*-weighted images were more sensitive to the difference in magnetization coefficient between normal brain tissue and the calcification.

The brain tissue under leptomeningeal hemangioma usually has limited brain atrophy. On CT and MRI, atrophy of the brain and enlargement of cisterns and sulci can be seen on the affected side. Brain atrophy often causes head asymmetry. Due to the lack of brain tissue growth in the lesion side, there is often skull is thickening, paranasal sinuses and mastoid air cell enlargement. Thickening of the skull on MRI shows high signal fat in widened diploic space of calvarium (Fig. 2.1.4). The midline structures are often significantly shifted to the lesion side. The white matter adjacent to the affected cerebral cortex shows low density on CT, and slight prolongation of T1 and T2 on MRI, which may represent glial hyperplasia in ischemic brain tissue [14].

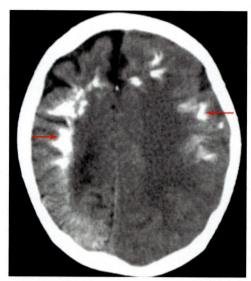

**FIGURE 2.1.3**   Axial CT reveals extensive bilateral cortical calcifications.

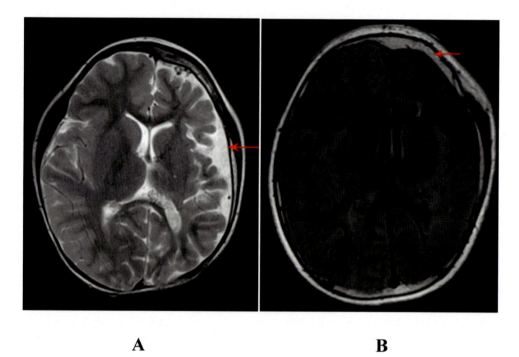

A                                                B

**FIGURE 2.1.4**   (A) Axial T2WI shows atrophy of the left hemisphere; (B) Axial T2-FLAIR shows the thickened overlying calvarium.

## 4. Skin

Impaired endothelial cell differentiation of dermal and subcutaneous tissue vessels and progressive expansion of immature small veins lead to congenital vascular malformations. The malformed vasculature consists of loosely arranged dilated thin-walled vessels, which cause the specific PWS when the skin vessels are involved. The vascular MAPK and/or PI3K signaling pathways are inextricably linked to normal development of the human embryo, and abnormalities in this signaling pathway not only lead to SWS and PWS but are also associated with disease progression; meanwhile, studies also have shown that somatic activation mutations in GNAQ resulted in PWS [6,8,15]. Specifically, the positive rate of GNAQ mutation in patients with PWS involving upper facial regions such as forehead, eyebrows, and upper eyelids is significantly higher than that in patients with PWS involving other facial regions. The positive rate of GNAQ mutation in patients with PWS in one or two facial regions is lower than that in patients with PWS in the upper, mid, and lower facial regions [16].

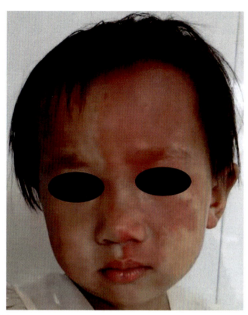

**FIGURE 2.1.5**  Photograph shows the classic port-wine stain of SWS.

It is estimated that the prevalence of PWS is 3—5 children per 1000 live births with no gender preference. The most common sites of PWS are the face, neck, trunk, and limbs (Fig. 2.1.5). Most facial PWS (~90%) are distributed along the unilateral trigeminal ganglia. 15%—20% of the patients with facial PWS involving the ophthalmic branch of the trigeminal nerve are at risk of suffering from SWS. In addition, these patients have a 50% increased risk of developing glaucoma, which almost always involves the eye on the same side as the facial PWS. In SWS, PWS appears at birth and persists throughout life [15], mostly unilaterally, occasionally bilaterally, with a similar chance of appearing on both sides, and can involve most or entire face. At first, the lesion is pink. With advancement in age, there is formation of vascular nodules which may became uplifted and thickened, and the nodules may be prone to spontaneous bleeding. Sometimes, multiple small red to purple nodules appear on the surface, making the lesions appear as cobblestone pattern [15,17,18]. Vascular cell proliferation caused by developmental disorders is the main phenotype in the pathogenesis of PWS. Vascular dilatation is a secondary abnormality, which occurs in the process of malformation development. Hyperactivity and proliferation of endothelial cells, pericytes and fibroblasts were observed in proliferative and nodular PWS [15,19].

## 5. Eyes

The two most common ocular diseases in SWS patients are glaucoma, which occurs in 30%—70% of patients, and choroidal hemangioma, which occurs in 20%—70% of patients [20,21]. Eye involvement can be characterized by increased conjunctival vessels, accompanied by ocular enlargement, and/or increased tearing in the affected eyes [22]. There are two favored ages for glaucoma in patients with SWS. Congenital glaucoma is characterized by early onset, affecting about 60% of patients and late-onset glaucoma affects 40% of SWS patients, occurring mainly in childhood and adolescence [23]. The most important cause of early-onset glaucoma is anterior chamber angle malformation, but in late-onset patients, elevated extrascleral venous pressure may be a more important cause. In addition, the immunopathogenesis and disease progression of glaucoma in SWS patients are associated with overexpression of inflammatory cytokines such as IL-7, mil-1a, and TNF-a [9,24].

Choroidal hemangioma is commonly seen in the eye of PWS involving the upper eyelid, which can be divided into limited and diffuse clinically, but diffuse choroidal hemangioma is more common in patients with SWS. Choroidal hemangiomas can lead to choroidal thickening, impaired vision, exudative retinal detachment, and macular edema and have the potential to increase the risk of glaucoma development. Imaging diagnosis of choroidal includes fundus fluorescein angiography, indocyanine green choroidal angiography, and enhanced depth imaging spectral domain optical coherence tomography, ultrasound, CT, and MRI [21,24,25]. On CT and MRI, SWS patients show giant or slightly larger glove; MRI can also show choroidal hemangiomas (Fig. 2.1.6). Choroidal hemangiomas are associated with bilateral onset and facial involvement, but not do not correlate with the size of intracranial leptomeningeal hemangioma. In T1WI,

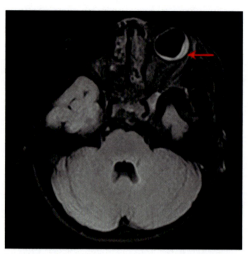

**FIGURE 2.1.6** Axial T2-FLAIR in a patient with SWS shows a diffuse choroidal angioma, seen here as a thick crescent around the posterior segment of the globe.

choroidal hemangioma shows thickening of the posterior wall of the eyeball. T2-weighted images show crescentic hyperintensity in the posterior wall of the globe, while fat-suppressed postcontrast MRI shows crescentic enhancement in the posterior wall of eyeball [17].

## 6. Endocrine system

In SWS patients, the frequency of a series of endocrine abnormalities is increased compared to the general population, including growth hormone deficiency and central hypothyroidism. The incidence of growth hormone deficiency is 18 times higher than the general population. Since the hypothalamic pituitary axis is normal on neuroimaging, the reason for this is not clear. IGF-1 level can be used to screen if there is growth hormone deficiency in children over 2 years old. If IGF-1 level is abnormally low, formal evaluation is needed to determine the diagnosis. Occasionally, central hypothyroidism or primary hypothyroidism may occur. Although it is related to the use of anticonvulsants, considering the increase of central growth hormone deficiency in SWS patients, hypothyroidism may be caused by hypothalamic—pituitary axis dysfunction related to the disease process. When suspected in the setting of excessive weight gain, dry skin, sleepiness, constipation, and other symptoms, one should consider testing of free T4 blood levels [2,26].

## References

[1] Cho S, Maharathi B, Ball KL, et al. Sturge—Weber syndrome patient registry: delayed diagnosis and poor seizure control. J Pediatr 2019;215:158—163.e156.

[2] Comi AM. Sturge—Weber syndrome. Handb Clin Neurol 2015;132:157—68.

[3] Ho TH, Yang FC, Lin JC, et al. Sturge—Weber syndrome. Qjm 2019;112(4):299.

[4] De la Torre AJ, Luat AF, Juhász C, et al. A multidisciplinary consensus for clinical care and research needs for Sturge—Weber syndrome. Pediatr Neurol 2018;84:11—20.

[5] Fjær R, Marciniak K, Sundnes O, et al. A novel somatic mutation in GNB2 provides new insights to the pathogenesis of Sturge—Weber syndrome. Hum Mol Genet 2021.

[6] Shirley MD, Tang H, Gallione CJ, et al. Sturge—Weber syndrome and port-wine stains caused by somatic mutation in GNAQ. N Engl J Med 2013;368(21):1971—9.

[7] Comi AM. Presentation, diagnosis, pathophysiology, and treatment of the neurological features of Sturge—Weber syndrome. Neurologist 2011;17(4):179—84.

[8] Maslin JS, Dorairaj SK, Ritch R. Sturge—Weber syndrome (*Encephalotrigeminal angiomatosis*): recent advances and future challenges. Asia Pac J Ophthalmol 2014;3(6):361—7.

[9] Bianchi F, Auricchio AM, Battaglia DI, et al. Sturge—Weber syndrome: an update on the relevant issues for neurosurgeons. Childs Nerv Syst 2020;36(10):2553—70.

[10] Sabeti S, Ball KL, Bhattacharya SK, et al. Consensus statement for the management and treatment of Sturge—Weber syndrome: neurology, neuroimaging, and ophthalmology recommendations. Pediatr Neurol 2021;121:59—66.

[11] Luat AF, Juhász C, Loeb JA, et al. Neurological complications of Sturge−Weber syndrome: current status and unmet needs. Pediatr Neurol 2019;98:31−8.

[12] Zallmann M, Leventer RJ, Mackay MT, et al. Screening for Sturge−Weber syndrome: a state-of-the-art review. Pediatr Dermatol 2018;35(1):30−42.

[13] Muralidharan V, Failla G, Travali M, et al. Isolated leptomeningeal angiomatosis in the sixth decade of life, an adulthood variant of Sturge Weber syndrome (type III): role of advanced magnetic resonance imaging and digital subtraction angiography in diagnosis. BMC Neurol 2020;20(1):366.

[14] Li YM, Liu C, Cong TX, et al. Clinical and neuroimaging analysis of 24 cases of Sturge−Weber syndrome. Sichuan Da Xue Xue Bao Yi Xue Ban 2020;51(4):562−6.

[15] Nguyen V, Hochman M, Mihm Jr MC, et al. The pathogenesis of port wine stain and Sturge Weber syndrome: complex interactions between genetic alterations and aberrant MAPK and PI3K activation. Int J Mol Sci 2019;20(9).

[16] Lee KT, Park JE, Eom Y, et al. Phenotypic association of presence of a somatic GNAQ mutation with port-wine stain distribution in capillary malformation. Head Neck 2019;41(12):4143−50.

[17] Baselga E. Sturge−Weber syndrome. Semin Cutan Med Surg 2004;23(2):87−98.

[18] Singh AK, Keenaghan M. Sturge−Weber syndrome. StatPearls Publishing Copyright © 2021. StatPearls Publishing LLC; 2021.

[19] Gao L, Yin R, Wang H, et al. Ultrastructural characterization of hyperactive endothelial cells, pericytes and fibroblasts in hypertrophic and nodular port-wine stain lesions. Br J Dermatol 2017;177(4):e105−8.

[20] Arora KS, Quigley HA, Comi AM, et al. Increased choroidal thickness in patients with Sturge−Weber syndrome. JAMA Ophthalmol 2013;131(9):1216−9.

[21] Mantelli F, Bruscolini A, La Cava M, et al. Ocular manifestations of Sturge−Weber syndrome: pathogenesis, diagnosis, and management. Clin Ophthalmol 2016;10:871−8.

[22] Sullivan T, Clarke M, Morin J. The ocular manifestations of the Sturge−Weber syndrome. J Pediatr Ophthalmol Strabismus 1992;29(6):349−56.

[23] Sujansky E, Conradi S. Sturge−Weber syndrome: age of onset of seizures and glaucoma and the prognosis for affected children. J Child Neurol 1995;10(1):49−58.

[24] Silverstein M, Salvin J. Ocular manifestations of Sturge−Weber syndrome. Curr Opin Ophthalmol 2019;30(5):301−5.

[25] Sinawat S, Auvichayapat N, Auvichayapat P, et al. 12-year retrospective study of Sturge−Weber syndrome and literature review. J Med Assoc Thai 2014;97(7):742−50.

[26] Sudarsanam A, Ardern-Holmes SL. Sturge−Weber syndrome: from the past to the present. Eur J Paediatr Neurol 2014;18(3):257−66.

Chapter 2.2

# Multisystem effects of neurofibromatosis

## 1. Introduction

NF, which includes three subtypes, neurofibromatosis type 1 (NF1) and neurofibromatosis type 2 (NF2) and schwannomatosis, is a group of dominantly inherited disorders with very low incidence and no racial preference. There is one person suffering from NF1 in every 3000 individuals all over the world [1]. The characteristic lesions of NF1 are skin and bone lesions; at the same time, NF1 is also associated with benign or malignant tumors of the CNS, and the most common tumor is benign neurofibroma [2]. Because of increased incidence of malignant tumors, heart attack, and stroke, the life expectancy of healthy people in general population is about 15 years longer than that of NF1 patients. The incidence of NF2 is about 1/30,000, which is 10 times lower than the incidence of NF1 [3]. NF is a tumor predisposition syndrome that predisposes patients to the development of a variety of benign and malignant tumors affecting multiple organs and systems [4]. Here, we will focus on the multisystem effects of NF (Table 2.2.1).

## 2. Pathology

Neurofibromin is a GTPase-activating protein that accelerates RAS signaling inactivation and is encoded by the NF1 gene. Mutations in the NF1 gene result in the deletion of neurofibromin, which leads to continuous activation of the RAS pathway. Active RAS transmits its growth-promoting signals through small protein cascades which are phosphorylated by kinases leading to successive activation of RAS downstream effector molecules, and eventually leading to the occurrence of benign and malignant tumors [4,5].

The NF2 gene on chromosome 22 primarily encodes the cytoskeletal protein merlin. NF2 protein acts as a tumor suppressor, and the double-gene allelic deletion of NF2 gene leads to the occurrence of tumors [5]. Merlin connects cytoskeletal proteins and membrane-associated proteins as an interface with the extracellular environment [6]. Although the mechanism of merlin deficiency leading to tumorigenesis is not completely clear, many studies have shown that merlin can bind to multiple effectors to regulate cell growth through different signaling pathways [4,5].

## 3. Central nervous system

Optic gliomas (OPGs) are seen in 15%—20% of NF1 patients and are the most common brain tumors in these patients [7]. OPG tend to occur in children, especially those younger than 7 years of age. Optic nerve and optic chiasma are the most prone sites for tumors, but they can also occur in the entire optic nerve pathway. From the pathological point of view,

**TABLE 2.2.1 The effect of Neurofibromatosis in each system.**

| Impact system | Major disease | Image findings | |
|---|---|---|---|
| Central nervous system | Tumor | NF1 | NF2 |
| | | Optic gliomas, brain stem glioma | Vestibular schwannoma, meningioma, ventricular meningiomas |
| Musculoskeletal system | Skeletal deformities | Scoliosis, long bone dysplasia, plexiform neurofibroma | |
| Respiratory system | Interstitial lung lesion | Parenchymal cyst, linear, lattice-like or ground-glass shadows | |
| Skin | Skin pigmentation, dermal neurofibromas | Skin pigmentation, dermal neurofibromas | |

OPGs are hairy cell astrocytomas and have a low likelihood of developing into highly malignant tumors. The diagnosis of OPG requires MRI, where OPG appears as thickening and swelling of the optic nerve with significant enhancement on postcontrast MRI. However, the absence of lesion on MRI does not mean that late onset OPG will not appear. Early age of onset, optic nerve involvement, and female gender are all risk factors for visual loss in NF1 patients with OPG [4]. Annual vision assessments are recommended for children to monitor changes in vision [8]. Because OPG occurs in the optic chiasm in most of the patients with NF1, precocious puberty signs appear in a small number of NF1 patients with OPG. Once precocious puberty occurs in children, the hypothalamus—pituitary—gonadal endocrine axis should be examined first [9], to investigate for possible optic chiasma—hypothalamus glioma and then it should be further monitored.

Brain stem glioma (BSG) ranks second most common among the brain tumors of NF1 patients [10]. The average onset age of BSG in NF1 patients is 7.2 years, which is older than that of OPG. In terms of histopathology, BDG is also a hairy cell astrocytoma that can grow anywhere in the brainstem, but in 66% of patients it is located in the midbrain and medulla oblongata. BDG is asymptomatic in 54% of NF1 patients, most of whom are occasionally detected during OPG monitoring [4,11]. Low-grade gliomas occur primarily in childhood and do not progress to malignancy, but less than 1% of adults also develop low-grade gliomas and have a high chance of developing malignancy. If adult patients with NF1 have neurological symptoms, imaging examination should be carried out in a timely fashion [4], because the risk of high-risk tumors in adult patients with NF1 is 10—50 times higher than that in the general population [12].

Vestibular schwannoma (VS) (Fig. 2.2.1) is the most common brain tumor in NF2 patients, which can be detected in 90%—95% of NF2 patients. Due to their special growth position, NF2 patients with VSs will have hearing impairment, tinnitus, or loss of balance. The best imaging method to diagnose VSs is high-resolution—enhanced MRI through the internal auditory canal plane. Considering the patient's age, tumor size, hearing loss, and other comprehensive situations, treatment options may include surgery or radiation therapy [4].

Meningioma (Fig. 2.2.2) ranked second in the common brain tumors of NF2 patients. Intracranial meningiomas are detected in 45%—50% of NF2 patients, while spinal meningiomas are detected in 20% of NF2 patients. Meningiomas in NF2 patients are predominantly multiple and occur at a younger age than that in the general population. Whether and what symptoms are present in NF2 patients with meningiomas is related to the size and location of the meningioma. Postcontrast MRI is the best imaging modality to diagnose meningiomas, and the tumor shows significant uniform enhancement [4]. The vast majority of meningiomas can be completely resected, and tumors that cannot be completely resected can be treated by radiotherapy as a supplemental method [13].

Ventricular meningiomas occur in 33%—35% of patients with NF2 and most often involve the cervical medulla or cervicomedullary junction (86%). Spinal cord tumors in patients with NF2 are usually asymptomatic [14], with tumor-related symptoms such as sensory disturbances, pain, or weakness occurring in less than 20% of patients. Ventricular meningiomas, presenting as hypointense or isointense on T1-weighted sequences and hyperintense on T2-weighted sequences, demonstrate enhancement on postcontrast MRI. Surgical treatment is usually effective for ventricular meningiomas in patients with NF2. Radiotherapy or chemotherapy is also an alternative treatment for patients with partial resection or recurrence of tumors [4,15].

## 4. Musculoskeletal system

NF1 is associated with osseous lesions including scoliosis, dysplasia of the sphenoid wing, and thinning of long bones [16]. The main spinal changes in NF1 patients are scoliosis and vertebral abnormalities [17]. Scoliosis, the most prevalent spinal deformity in NF1 patients (Fig. 2.2.3), is seen in 30%—70% of NF1 patients and occurs mainly in the thorax, either as a primary result of bony dysplasia, secondary to chronic compression, or a combination of both [18,19]. Scoliosis in NF1 patients mainly includes dystrophic and nondystrophic types, with the former being the most common. Dystrophic scoliosis most commonly involves 4—6 spinal segments and is characterized by short segmental involvement and sharp scoliosis angles [20]. Dystrophic scoliosis should be treated early because of its rapid progression. Dystrophic features such as scalloped or wedge-shaped vertebrae, pencil-shaped ribs, and fusiform transverse processes predict that the deformity will continue to progress in the future [21]. In nondystrophic scoliosis, the presentation is not only similar to that of adolescent idiopathic scoliosis, but the treatment modality is also the same [20].

Long bone dysplasia is probably the most typical skeletal abnormality in children with NF1. The characteristic change is the anterolateral bowing of the tibia centered at the junction of the middle and lower 1/3 of the tibia. In some patients with NF1, fractures occur in the bowing tibia and the pathologic fracture is located in the tibial bowing region [22,23]. Fractures of bent tibia are common in early childhood, most of which cannot heal normally, and eventually develop into nonunion or pseudojoint. This series of lesions involved almost only unilateral tibia [24]. The typical X-ray changes of tibial dysplasia in NF1 patients were cortical thickening with narrowing of the bone marrow cavity, but a small number of

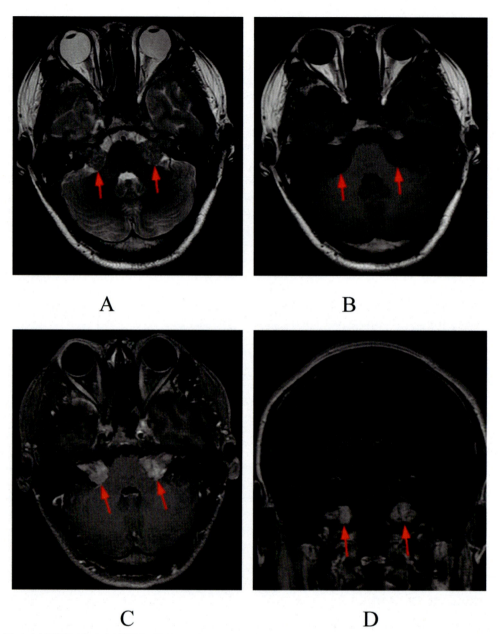

A

B

C

D

**FIGURE 2.2.1** (A) Axial T1WI. (B) Axial T2WI. (C) Postcontrast axial T1WI. (D) Postcontrast coronal T1WI. (A–D) Show bilateral masses in the pontocerebellar horn region protrude into the internal auditory tract, resulting in enlargement of the internal auditory tract. Postoperative pathology confirmed schwannoma.

patients showed opposite manifestations, namely cortical thinning. Other X-ray manifestations also included widening of the medullary canal with tubular defects, cystic lesions, and contraction of long bones with dysplasia. Clinically, once the anterolateral bowing of the tibia is found, it should be referred to the department of orthopedics as soon as possible to avoid progression to fracture and pseudoarthrosis [23].

Plexiform neurofibroma (pFN) is composed of multiple cells, which is one of the markers of NF and is associated with chronic pain (Fig. 2.2.4). pFN can be found in up to half of patients with NF1 and tumors vary in size, ranging from limited to diffuse lesions [25,26]. pNF is common in children and progresses rapidly during this period, with tumor volume growth rates of >20% per year. The prone sites of tumors were paravertebral, head, neck, and limbs. Although they were benign tumors, they could show invasive growth. pFN tumors typically present as masses with hypointense T1 signal and hyperintense T2 signal. The main treatment for pFN is still surgery. Early surgery may help to avoid complications, but complete removal of the tumor is challenging and therefore the tumor has a high recurrence rate [17,27,28].

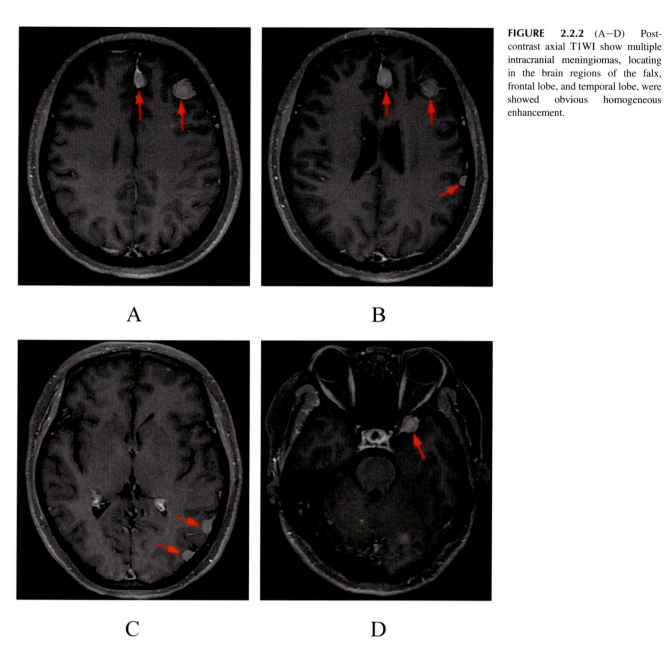

**FIGURE 2.2.2** (A—D) Post-contrast axial T1WI show multiple intracranial meningiomas, locating in the brain regions of the falx, frontal lobe, and temporal lobe, were showed obvious homogeneous enhancement.

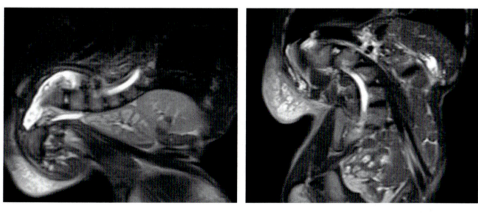

**FIGURE 2.2.3** (A and B) Coronal T2WI show "S-shaped scoliosis" of thoracolumbar in an NF1 patient.

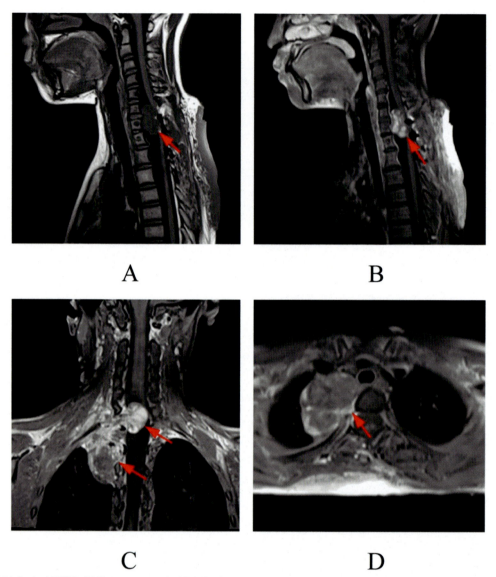

**FIGURE 2.2.4** (A) Sagittal T1WI. (B) Postcontrast sagittal T1WI. (C) Postcontrast coronal T1WI. (D) Postcontrast axial T1WI. (A—D) reveal mass on the right side and in the spinal canal at the level of the sixth cervical vertebra—third thoracic vertebra, partially dumbbell-shaped, with multiple enlarged intervertebral foramina.

## 5. Respiratory system

Previous studies have found that 5 out of 100 individuals develop parenchymal neurofibromas [29]. Due to the different sizes and locations of lesions, they lead to a series of nonspecific respiratory symptoms from asymptomatic to cough, dyspnea, pneumonia. Lung biopsy in patients with NF1 patients reveals parenchymal cysts and centrilobular nodules without clear walls [30] (Fig. 2.2.5). These lesions are mostly located in the upper lung [31], and resemble centrilobular emphysema as the disease progresses, often leading to misdiagnosis. These lesions are confusing because they are often reported in patients with smoking. However, a study found that all six asymptomatic nonsmoking NF1 patients had pulmonary cystic lesions in chest imaging examination [32]. In the late stage of the disease, these lesions will develop into chronic respiratory failure or even more serious complications such as spontaneous pneumothorax or pulmonary hypertension [30,33].

Since 1963, studies have been conducted to demonstrate the association between NF1 and infiltrative diseases that can progress to pulmonary fibrosis, such as cystic lung lesions and emphysema [30]. The incidence of these diseases is as high

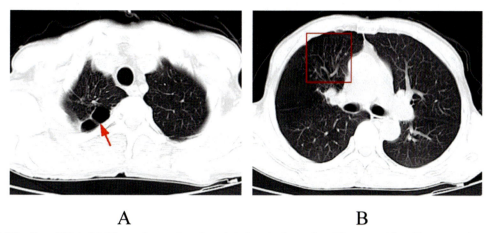

**FIGURE 2.2.5**   (A and B) Axial CT reveal parenchymal cystic lesions (*red arrow*) and linear opacities of lung parenchyma (*red box*).

as 20% [29]. Patients affected by interstitial lung injury tend to present clinically with dyspnea, but a small percentage of patients are asymptomatic. On CT, these diseases appear as linear, lattice-like, or ground-glass shadows; typical honey-comb imaging appearance is rare [29,30,34]. But whether there are NF1-specific interstitial lesions is uncertain. Some investigators have linked smoking to this disease, arguing that mutations in the NF1 gene resulted in a failure to regulate proliferation or apoptosis in lung basal cells [35]. Some researchers found that smoking had no relationship with interstitial lesions, so they attributed lung lesions to the mutation of NF1 gene [31,32]. Although there is no consistent conclusion, it is necessary to encourage NF1 patients to quit smoking. On the one hand, macrophage alveolar inflammation caused by smoking can lead to the formation of pulmonary cysts or fibrosis-related lesions in NF1 patients. On the other hand, NF1 may increase the sensitivity of the lung to smoking. In addition, a large number of nerve growth factors were found in the serum of NF1 patients, which may promote fibroblast proliferation and lung disease [30].

Pulmonary hypertension associated with NF1 (PH-NF1) is a rare complication that belongs to the hypertension group 5. PH-NF1 is more common in women, and there is approximately a lag of 45 years between the occurrence of pulmonary hypertension and the diagnosis of NF1, so that PH-NF1 tends to present in the late stage of the disease. Patients with NF1 may have interstitial lung disease, but the degree of pulmonary hypertension is disproportionate to lung lesions. Some patients with mild lung lesions develop severe pulmonary hypertension, which supports the hypothesis that there may be specific pulmonary vascular diseases. PH-NF1 patients have symptoms of dyspnea and right heart failure; in addition, about 3/4 of them have New York cardiac class III or IV with severe limitation of exercise function. There is lack of data on the efficacy of treatment for this group of patients, so it is recommended that patients with PH-NF1 are evaluated at a specialized hypertension center and undergo early lung transplantation [30,36].

## 6. Skin

The most common feature recognized by NF1 is skin pigmentation. The café-au-lait spot is found in 95% of patients and is the first clinical manifestation which can assist in the diagnosis of NF1. Approximately 70% of patients with NF1 develop freckles about 1−2 mm in size in skin folds, such as the axillae and groin, between 3 and 5 years of age. This is generally known as Crowe's sign and is one of the diagnostic criteria for NF1 [30,37]. Up to 50% of NF1 patients will have a flat spot lighter than the surrounding skin color, and friction cannot make it red, which is termed as anemic mole [38]. Juvenile xanthogranulomas only appear in a small number of NF1 patients, which are characterized by small yellow nodules and can spontaneously decompose [39]. In addition, some patients may develop pruritus of the skin.

Dermal neurofibromas increase with age and mostly appear in adolescence. Multiple separate benign neurofibromas are present in the dermis in approximately 95% of patients. Cutaneous neurofibromas can be found in almost all adult NF1 patients, and it involves the skin in a certain sequence, starting with the trunk and then progressing to the extremities, neck, and face. The number of neurofibromas varies from person to person, but they are essentially lavender nodules that protrude from the skin surface or are invaginated beneath the skin [37] (Fig. 2.2.6).

**FIGURE 2.2.6** (A and B) Axial CT shows multiple cutaneous neurofibromas located in the abdomen and back.

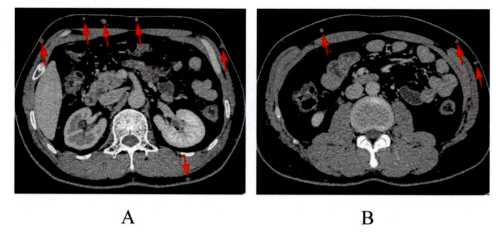

A                    B

# References

[1] Bakker AC, La Rosa S, Sherman LS, et al. Neurofibromatosis as a gateway to better treatment for a variety of malignancies. Prog Neurobiol 2017;152:149–65.

[2] Mirza T, Majeed MH. Neurofibromatosis and psychosis: coincidence or co-occurrence? Aust N Z J Psychiatry 2019;53(6):585–6.

[3] Farschtschi S, Mautner VF, McLean ACL, et al. The neurofibromatoses. Dtsch Arztebl Int 2020;117(20):354–60.

[4] Campian J, Gutmann DH. CNS tumors in neurofibromatosis. J Clin Oncol 2017;35(21):2378–85.

[5] Plotkin S, RWick A. Neurofibromatosis and schwannomatosis. Semin Neurol 2018;38(1):73–85.

[6] McClatchey A, IGiovannini M. Membrane organization and tumorigenesis—the NF2 tumor suppressor, Merlin. Genes Dev 2005;19(19):2265–77.

[7] Listernick R, Charrow J, Greenwald MJ, et al. Optic gliomas in children with neurofibromatosis type 1. J Pediatr 1989;114(5):788–92.

[8] Fisher MJ, Avery RA, Allen JC, et al. Functional outcome measures for NF1-associated optic pathway glioma clinical trials. Neurology 2013;81(21 Suppl. 1):S15–24.

[9] Habiby R, Silverman B, Listernick R, et al. Precocious puberty in children with neurofibromatosis type 1. J Pediatr 1995;126(3):364–7.

[10] Ullrich NJ, Raja AI, Irons MB, et al. Brainstem lesions in neurofibromatosis type 1. Neurosurgery 2007;61(4):762–6 (discussion 766–767).

[11] Mahdi J, Shah AC, Sato A, et al. A multi-institutional study of brainstem gliomas in children with neurofibromatosis type 1. Neurology 2017;88(16):1584–9.

[12] Taddei M, Erbetta A, Esposito S, et al. Brain tumors in NF1 children: influence on neurocognitive and behavioral outcome. Cancers 2019;11(11).

[13] Wentworth S, Pinn M, Bourland JD, et al. Clinical experience with radiation therapy in the management of neurofibromatosis-associated central nervous system tumors. Int J Radiat Oncol Biol Phys 2009;73(1):208–13.

[14] Plotkin SR, O'Donnell CC, Curry WT, et al. Spinal ependymomas in neurofibromatosis Type 2: a retrospective analysis of 55 patients. J Neurosurg Spine 2011;14(4):543–7.

[15] Kresbach C, Dorostkar MM, Suwala AK, et al. Neurofibromatosis type 2 predisposes to ependymomas of various localization, histology, and molecular subtype. Acta Neuropathol 2021;141(6):971–4.

[16] Hernández-Martín A, Duat-Rodríguez A. An update on neurofibromatosis type 1: not just Café-au-lait spots, freckling, and neurofibromas. An update. Part I. Dermatological clinical criteria diagnostic of the disease. Actas Dermosifiliogr 2016;107(6):454–64.

[17] Shofty B, Barzilai O, Khashan M, et al. Spinal manifestations of Neurofibromatosis type 1. Childs Nerv Syst 2020;36(10):2401–8.

[18] Well L, Careddu A, Stark M, et al. Phenotyping spinal abnormalities in patients with Neurofibromatosis type 1 using whole-body MRI. Sci Rep 2021;11(1):16889.

[19] Khong PL, Goh WH, Wong VC, et al. MR imaging of spinal tumors in children with neurofibromatosis 1. AJR Am J Roentgenol 2003;180(2):413–7.

[20] Tsirikos AI, Saifuddin A, Noordeen MH. Spinal deformity in neurofibromatosis type-1: diagnosis and treatment. Eur Spine J 2005;14(5):427–39.

[21] Durrani AA, Crawford AH, Chouhdry SN, et al. Modulation of spinal deformities in patients with neurofibromatosis type 1. Spine 2000;25(1):69–75.

[22] Crawford AH, Schorry EK. Neurofibromatosis update. J Pediatr Orthop 2006;26(3):413–23.

[23] Stevenson DA, Viskochil DH, Schorry EK, et al. The use of anterolateral bowing of the lower leg in the diagnostic criteria for neurofibromatosis type 1. Genet Med 2007;9(7):409–12.

[24] Stevenson DA, Hanson H, Stevens A, et al. Quantitative ultrasound and tibial dysplasia in neurofibromatosis type 1. J Clin Densitom 2018;21(2):179–84.

[25] Ferner RE, Huson SM, Thomas N, et al. Guidelines for the diagnosis and management of individuals with neurofibromatosis 1. J Med Genet 2007;44(2):81–8.

[26] Le LQ, Liu C, Shipman T, et al. Susceptible stages in Schwann cells for NF1-associated plexiform neurofibroma development. Cancer Res 2011;71(13):4686−95.

[27] Well L, Jaeger A, Kehrer-Sawatzki H, et al. The effect of pregnancy on growth-dynamics of neurofibromas in Neurofibromatosis type 1. PLoS One 2020;15(4):e0232031.

[28] Blakeley JO, Plotkin SR. Therapeutic advances for the tumors associated with neurofibromatosis type 1, type 2, and schwannomatosis. Neuro Oncol 2016;18(5):624−38.

[29] Ryu JH, Parambil JG, McGrann PS, et al. Lack of evidence for an association between neurofibromatosis and pulmonary fibrosis. Chest 2005;128(4):2381−6.

[30] Reviron-Rabec L, Girerd B, Seferian A, et al. Pulmonary complications of type 1 neurofibromatosis. Rev Mal Respir 2016;33(6):460−73.

[31] Zamora AC, Collard HR, Wolters PJ, et al. Neurofibromatosis-associated lung disease: a case series and literature review. Eur Respir J 2007;29(1):210−4.

[32] Oikonomou A, Vadikolias K, Birbilis T, et al. HRCT findings in the lungs of non-smokers with neurofibromatosis. Eur J Radiol 2011;80(3):e520−3.

[33] Ayed Della S, Kotti A, Ben Sik Ali H, et al. Spontaneous pneumothorax and Recklinghausen's disease: a case report. Rev Pneumol Clin 2012;68(3):202−4.

[34] Shino MY, Rabbani S, Belperio JA, et al. Neurofibromatosis-associated diffuse lung disease: case report. Semin Respir Crit Care Med 2012;33(5):572−5.

[35] Katzenstein AL. Smoking-related interstitial fibrosis (SRIF): pathologic findings and distinction from other chronic fibrosing lung diseases. J Clin Pathol 2013;66(10):882−7.

[36] Jutant EM, Girerd B, Jaïs X, et al. Pulmonary hypertension associated with neurofibromatosis type 1. Eur Respir Rev 2018;27(149).

[37] Miller DT, Freedenberg D, Schorry E, et al. Health Supervision for children with neurofibromatosis type 1. Pediatrics 2019;143(5).

[38] Marque M, Roubertie A, Jaussent A, et al. Nevus anemicus in neurofibromatosis type 1: a potential new diagnostic criterion. J Am Acad Dermatol 2013;69(5):768−75.

[39] Ferrari F, Masurel A, Olivier-Faivre L, et al. Juvenile xanthogranuloma and nevus anemicus in the diagnosis of neurofibromatosis type 1. JAMA Dermatol 2014;150(1):42−6.

Chapter 2.3

# Multisystem effects of tuberous sclerosis complex

## 1. Introduction

The mutation of the tumor suppressor gene TSC 1 or TSC 2 results in TSC, which is an autosomal inherited disease and is known for the discovery of thick, hard and pale gyrus in the brain of postmortem patients. One out of every 6000 to 10,000 people in the whole world suffers from TSC, and there is no racial and gender difference. Both children and adults can develop the disease, and it currently affects approximately two million people worldwide [1—3]. The main clinical manifestations of TSC, which affects multiple organ systems (Table 2.3.1), can be divided into two categories. One is the tumor which can occur in multiple organs, affecting the brain, lung, heart, kidney, and skin. Another represents neuro-psychiatric diseases, such as epilepsy, cognitive impairment, and autism spectrum disorders. Here we will describe the multisystem effects of tuberous sclerosis [4] (Fig. 2.3.1).

**TABLE 2.3.1** The effects of TSC in each system.

| Impact system | Major diseases | Images finding |
|---|---|---|
| Central nervous system | Epilepsy, subependymal nodule, intellectual disability, autism spectrum disorder. | Cortical nodule, subependymal giant cell astrocytoma |
| Respiratory system | Lymphangioleiomyomatosis, multifocal micronodular pneumocyte hyperplasia. | Cystic destruction of the lungs |
| Circulatory system | Cardiac rhabdomyoma. | Cardiac rhabdomyoma |
| Urinary system | Angiomyolipoma, simple multiple cysts, polycystic kidney disease, renal cell carcinoma. | Angiomyolipoma, cysts, renal cell carcinoma |
| Skin | Fibroma, focal hypopigmentation. | |

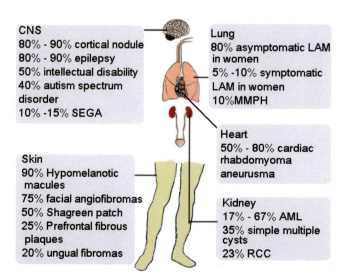

**FIGURE 2.3.1** Tuberous sclerosis complex results in multiple system impairment. *The picture was modified from Henske EP, Jóźwiak S, Kingswood JC, et al. Tuberous sclerosis complex. Nat Rev Dis Primers 2016;2:16035.*

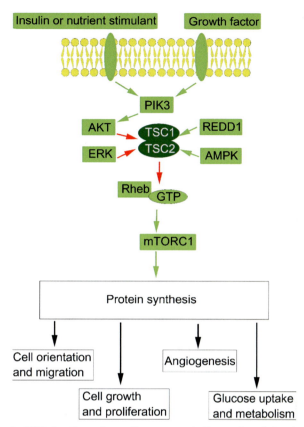

**FIGURE 2.3.2**  TSC1-TSC2 complex and mTOR signaling pathway. *Green arrows* indicate activated phosphorylation sites; *red arrows* show inhibition of phosphorylation sites. *This picture was modified from Curatolo P, Moavero R, de Vries PJ. Neurological and neuropsychiatric aspects of tuberous sclerosis complex. Lancet Neurol 2015;14(7):733—45.*

## 2. Pathogenesis

TSC is an inherited disease caused by gene mutation on chromosome 19 or chromosome 16. Hamartin (TSC1) protein is encoded by TSC gene 1, while tuberin (TSC2) protein is encoded by TSC gene 2. The mammalian target of rapamycin (mTOR) pathway (Fig. 2.3.2) is regulated by TSC1-TSC2 complex which is composed of TSC1 protein and TSC2 protein and plays the role of tumor suppressor. MTOR, as a serine/threonine kinase, is involved in regulating the metabolism, growth, and survival of cells in the body. Mutations of TSC1 gene or TSC2 gene cause mTOR complex 1 (mTORC1) and mTOR complex 2 (mTORC2) to fail to function normally, resulting in excessive activation of mTOR pathway. MTORC1 is primarily responsible for regulating the cell cycle and protein synthesis, while mTORC2 regulates the actin cytoskeleton. The downstream signaling pathway of mTOR pathway involves protein synthesis and metabolism, cell growth, and reproduction. mTOR signal imbalance leads to increased cell growth and proliferation [1,5].

## 3. Central nervous system

Cortical nodule is a form of neuroglial hamartoma, affecting 80%—90% of TSC patients [7]. In patients with tumor suppressor gene TSC1 mutation, the number and size of cortical nodules are greater [8]. Cortical nodules are more common in the supratentorial region, with a preference for the frontal lobe, but epileptogenic nodules are more common in the inferior parietal lobule and the central sulcus. Their number and distribution correlate with the severity of symptoms. Whether the seizures originate in the nodules themselves or in the perinodal cortex is still inconclusive. Histology is characterized by dysplastic neurons, giant cell changes, and laminar disorders. On macroscopic observation, the affected brain gyrus is thickened and bulging. On MRI, the nodules show low T1 signal and high T2/FLAIR signal [9—11].

The brain is the most severely affected organ system, leading to epilepsy, neuropsychiatric disorders, anxiety, and tumors [1,5]. Diseases of the nervous system are the main reason for the presentation of TSC patients. 80%—90% of TSC patients will have neurological-related clinical manifestations. Among them, epilepsy is the most universal neurological symptom, which occurs mostly within 1 year age. In contrast, in those children with large cortical malformations, epilepsy can present within 28 days of birth [1]. Early onset of epilepsy causes mental retardation, language abnormalities, and autism-related behaviors in up to 50% of patients [12]. Epilepsy in patients with TSC begins as partial seizures, and over

**FIGURE 2.3.3** Brain MRI of TSC patients. (A) Axial T1WI shows bilateral subependymal multiple nodules (*arrow*). (B) Postcontrast axial T1WI reveals bilateral subependymal multiple nodules without enhancement (*arrow*).

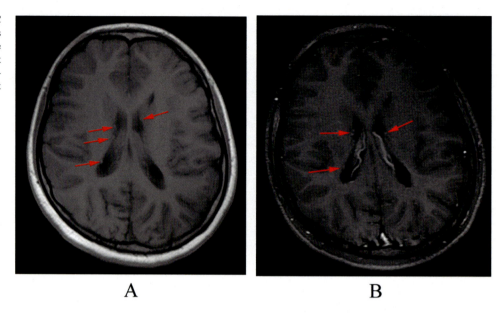

A                                        B

time, in some patients, infantile spasms develop [6]. And TSC patients are twice as likely to develop intractable or refractory epilepsy than general epilepsy patients [1].

Cognitive and neurobehavioral problems are universal in TSC patients. It is reported that half of TSC patients suffer from varying degrees of mental retardation, and as many as 40% of TSC patients have autism spectrum disorders. Specific cognitive impairments are common even in people with normal IQ. TSC patients with early onset epilepsy or refractory epilepsy are more likely to develop mental retardation. In addition, about 66% of TSC patients have mental health problems, among which anxiety is particularly common, affecting about 30%—60% of patients [1,13].

Subependymal giant cell astrocytoma (SEGA) (Fig. 2.3.3) can be found in 10%—15% of TSC patients. The incidence is common in adolescents, especially in those under 20 years of age, but the incidence in fetuses and infants is also reported [1]. For the definition of the disease, most investigators consider subependymal nodules larger than 1 cm in diameter, especially around Monroe's foramen, to be SEGA [14]. When SEGA causes obstructive hydrocephalus or drug-refractory epilepsy, neurosurgery is required to remove the obstruction or control the seizures [15].

## 4. Respiratory system

The main disease of the lungs in TSC patients is lymphangioleiomyomatosis (LAM), which can lead to lung tissue destruction that is manifested as multiple cystic lesions, pneumothorax, and celiac pleural effusion, resulting in shortness of breath, chest pain, and fatigue [1]. If female TSC patients suffering from (LAM), it indicates that TSC gene 1 and TSC gene 2 are both mutated [16]. Asymptomatic LAM mostly occurs in women and affects approximately 80% of female patients with TSC. 5%—10% of female patients with LAM have clinical manifestations which include respiratory failure in severe cases [1]. The development of LAM in perimenopausal period will accelerate and cases have been reported in which worsening shortness of breath or progressive pneumothorax can occur during pregnancy; therefore, estrogen and pregnancy are considered to aggravate the progress of the disease [17,18]. In male TSC patients, there are few patients with LAM confirmed by biopsy. Another lung manifestation of TSC patients is multifocal nodular pulmonary cell hyperplasia (MMPH), which is different from LAM as there is not gender preference. In general, MMPH affects 10% of TSC patients and most do not have clinical manifestations [1,19].

## 5. Circulatory system

Cardiac rhabdomyosarcoma is the most representative manifestation of the heart in patients with TSC, consisting of atypically enlarged myocytes, and is the earliest misshapen tumor to be identified. Cardiac rhabdomyosarcoma can be detected in the fetus or newborn and has an incidence of approximately 50%—80% in patients with TSC. Infants with cardiac rhabdomyosarcoma have a 96% chance of having a final diagnosis of TSC. Although all ventricles can be involved, cardiac rhabdomyosarcoma is most commonly found in the left ventricle. Larger lesions can interfere with blood flow or cause cardiac embolic disease, which clinically manifests as heart rate and conduction disturbances, nonimmune edema, and even death, and therefore sometimes requires surgical removal. However, in most patients, cardiac rhabdomyosarcomas are asymptomatic and resolve naturally with age. In patients over 10 years of age, cardiac rhabdomyosarcoma usually does not require treatment, but in children and infants under 10 years of age, cardiac rhabdomyoma ranks first among the causes of death. The main vascular abnormalities in patients with TSC are aneurysms, and the most frequently involved vessels are carotid artery, axillary artery and renal artery, and intracranial aneurysms occur rarely [20,21].

## 6. Urinary system

Renal damage in TSC patients begins in early childhood and worsens in adulthood [22]. The two most common lesions that can be detected in TSC patients are angiomyolipomas (AMLs) and cysts (Fig. 2.3.4), which can be found in childhood. AML

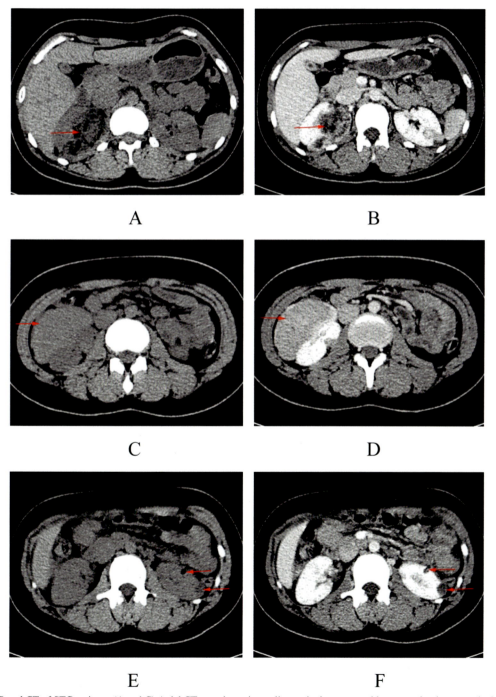

A          B

C          D

E          F

**FIGURE 2.3.4** Renal CT of STC patients. (A and C) Axial CT reveals angiomyolipoma in the upper and lower renal pole, respectively (*arrow*). (B and D) Contrast axial CT shows renal angiomyolipoma with varying degrees of enhancement. (E) Axial CT reveals cysts in left kidney. (F) Contrast axial CT shows cysts in left kidney without enhancement.

continues to develop in childhood and early adulthood, with detecting in approximately 17% of children at age 2 years, 65% at ages 9—14 years, and 67% in adulthood [1]. AMLs occur most frequently in the renal cortex, with bilateral multiple lesions seen in adult TSC patients. The clinical presentation is characterized by abdominal pain and impaired renal function, and spontaneous bleeding of AML occurs in severe cases [22]. Women are more likely to develop AMLs and will have a greater number of tumors [23]. The growth rate of AML varies with age, and accelerates after the age of 12 years [24].

Renal cysts can be found in 35% of TSC patients, especially multiple simple renal cysts [1]. Other renal manifestations of TSC patients include renal cell carcinoma (RCC), epithelioid change of AML, and oncocytoma [1]. RCC can present as multifocal and bilateral, affecting 23% of patients with TSC, and children can also be affected. In TSC patients, the proportion of women who develop it is twice as high as that of men [1,25]. In addition, focal and segmental glomerulosclerosis can also occur in TSC patients [1].

## 7. Skin and oral cavity

Almost all patients with TSC have skin manifestations that include facial angiofibromas, ungual fibromas, fibrous cephalic plaques, shagreen patches, and focal hypopigmented changes. These cutaneous manifestations are helpful in the clinical diagnosis of TSC patients [1]. Hypomelanotic macules usually appear at birth or infancy, which can be observed in 90% of TSC patients and represent an important feature of TSC [4,6]. Facial angiofibromas can be observed in 75% of patients with TSC, with a common onset age of 2—5 years. Facial angiofibromas are common in multiple cases, but mild cases can present as limited facial angiofibromas [4,6,26]. Facial angiofibromas usually require treatment and are of great concern to patients [1]. Prefrontal fibrous plaques are found in 25% of patients with TSC. In fact, fibrous plaques, which may be the most special skin manifestation of TSC, are more common on the forehead, but also occur in the face or scalp [4,6]. Ungual fibromas, occuring in about 20% of patients with TSC but up to 80% in older patients, refer to the growth of fiber around the nail. The uneven or orange peel-like surface of the skin is called the Shagreen patch, which occurs mainly on the low back and is seen in 50% of patients with TSC, with the onset mostly within 10 years of age [1,4].

Dental enamel pits and oral fibroma are the main oral manifestations of TSC patients. Almost all patients develop dental enamel pits and gingival fibromas, with an incidence of 20%—50%, which are the most common oral fibromas and are more common in adults. Fibromas can also develop in the buccal or lip mucosa and even in the tongue [1,4].

## References

[1] Henske EP, Jóźwiak S, Kingswood JC, et al. Tuberous sclerosis complex. Nat Rev Dis Primers 2016;2:16035.

[2] O'Callaghan FJ, Shiell AW, Osborne JP, et al. Prevalence of tuberous sclerosis estimated by capture-recapture analysis. Lancet 1998;351(9114):1490.

[3] Hallett L, Foster T, Liu Z, et al. Burden of disease and unmet needs in tuberous sclerosis complex with neurological manifestations: systematic review. Curr Med Res Opin 2011;27(8):1571—83.

[4] Northrup H, Krueger DA. Tuberous sclerosis complex diagnostic criteria update: recommendations of the 2012 international tuberous sclerosis complex consensus conference. Pediatr Neurol 2013;49(4):243—54.

[5] Salussolia CL, Klonowska K, Kwiatkowski DJ, et al. Genetic etiologies, diagnosis, and treatment of tuberous sclerosis complex. Annu Rev Genomics Hum Genet 2019;20:217—40.

[6] Curatolo P, Moavero R, de Vries PJ. Neurological and neuropsychiatric aspects of tuberous sclerosis complex. Lancet Neurol 2015;14(7):733—45.

[7] Cotter JA. An update on the central nervous system manifestations of tuberous sclerosis complex. Acta Neuropathol 2020;139(4):613—24.

[8] Overwater IE, Swenker R, van der Ende EL, et al. Genotype and brain pathology phenotype in children with tuberous sclerosis complex. Eur J Hum Genet 2016;24(12):1688—95.

[9] Ellingson BM, Hirata Y, Yogi A, et al. Topographical distribution of epileptogenic tubers in patients with tuberous sclerosis complex. J Child Neurol 2016;31(5):636—45.

[10] Lu DS, Karas PJ, Krueger DA, et al. Central nervous system manifestations of tuberous sclerosis complex. Am J Med Genet C Semin Med Genet 2018;178(3):291—8.

[11] Kalantari BN, Salamon N. Neuroimaging of tuberous sclerosis: spectrum of pathologic findings and frontiers in imaging. AJR Am J Roentgenol 2008;190(5):W304—9.

[12] Overwater IE, Bindels-de Heus K, Rietman AB, et al. Epilepsy in children with tuberous sclerosis complex: chance of remission and response to antiepileptic drugs. Epilepsia 2015;56(8):1239—45.

[13] Toldo I, Brasson V, Miscioscia M, et al. Tuberous sclerosis-associated neuropsychiatric disorders: a paediatric cohort study. Dev Med Child Neurol 2019;61(2):168—73.

[14] Roth J, Roach ES, Bartels U, et al. Subependymal giant cell astrocytoma: diagnosis, screening, and treatment. Recommendations from the international tuberous sclerosis complex consensus conference 2012. Pediatr Neurol 2013;49(6):439—44.

[15] Krueger DA, Northrup H. Tuberous sclerosis complex surveillance and management: recommendations of the 2012 international tuberous sclerosis complex consensus conference. Pediatr Neurol 2013;49(4):255−65.

[16] Gupta N, Henske EP. Pulmonary manifestations in tuberous sclerosis complex. Am J Med Genet C Semin Med Genet 2018;178(3):326−37.

[17] Yu JJ, Robb VA, Morrison TA, et al. Estrogen promotes the survival and pulmonary metastasis of tuberin-null cells. Proc Natl Acad Sci USA 2009;106(8):2635−40.

[18] Henske EP, McCormack FX. Lymphangioleiomyomatosis—a wolf in sheep's clothing. J Clin Invest 2012;122(11):3807−16.

[19] Hayashi T, Kumasaka T, Mitani K, et al. Loss of heterozygosity on tuberous sclerosis complex genes in multifocal micronodular pneumocyte hyperplasia. Mod Pathol 2010;23(9):1251−60.

[20] Volpi A, Sala G, Lesma E, et al. Tuberous sclerosis complex: new insights into clinical and therapeutic approach. J Nephrol 2019;32(3):355−63.

[21] Wataya-Kaneda M, Uemura M, Fujita K, et al. Tuberous sclerosis complex: recent advances in manifestations and therapy. Int J Urol 2017;24(9):681−91.

[22] Lam HC, Siroky BJ, Henske EP. Renal disease in tuberous sclerosis complex: pathogenesis and therapy. Nat Rev Nephrol 2018;14(11):704−16.

[23] Franz DN, Belousova E, Sparagana S, et al. Long-term use of everolimus in patients with tuberous sclerosis complex: final results from the EXIST-1 study. PLoS One 2016;11(6):e0158476.

[24] Robert A, Leroy V, Riquet A, et al. Renal involvement in tuberous sclerosis complex with emphasis on cystic lesions. Radiol Med 2016;121(5):402−8.

[25] Henske EP, Cornejo K, MWu CL. Renal cell carcinoma in tuberous sclerosis complex. Genes 2021;12(10).

[26] Józwiak S, Schwartz RA, Janniger CK, et al. Usefulness of diagnostic criteria of tuberous sclerosis complex in pediatric patients. J Child Neurol 2000;15(10):652−9.

Chapter 2.4

# Multisystem effects of von Hippel-Lindau syndrome

## 1. Introduction

von Hippel-Lindau syndrome (VHL) is a rare genetic disease involving multiple organs with multiple tumor syndrome as the main clinical manifestation. Mutations of the VHL gene located on chromosome 3P25-26 lead to VHL, which affects 1 in 45,000 people worldwide and can occur in individuals from infancy to 70 years, but the most common age of onset is 18—30 years, with an average onset age of 26 years, and there is no gender preference of VHL [1]. Obligate heterozygotes may be asymptomatic well into adulthood, but by age 60, the explicit rate was close to 90%. VHL syndrome is often manifested as multifamily, multiple systems, and multiple organs involvement (Table 2.4.1), mainly involving the brain and spinal cord, retina, pancreas, kidney, adrenal gland, epididymis, and other organs, showing the coexistence of benign and malignant tumors [2,3]. VHL syndrome can be divided into two types based on the presence or absence of pheochromocytoma. Without concomitant pheochromocytoma, VHL is classified as type 1, but if VHL combined with pheochromocytoma, it is classified as type 2. Type 2 VHL is further split into three different subtypes: type 2A, which is characterized by pheochromocytoma with CNS hemangioblastoma but without RCC, type 2B, which has all pheochromocytoma, CNS hemangioblastoma, and rRCC, and type 2C, which only has pheochromocytoma [1]. Here, we will introduce the multisystem effects of VHL syndrome.

## 2. Pathogenesis

The VHL gene is a tumor suppressor gene that plays a vital role in the regulation of angiogenesis and cell division [1]. By controlling cell division and angiogenesis, the VHL gene acts as tumor suppressor. VHL is caused by the mutation of VHL gene on chromosome 3p25, which prevents the normal synthesis of VHL protein (pVHL) or causes improper production of pVHL and impairs the normal function of pVHL and eventually leads to VHL [4]. Hypoxia-inducible factor (HIF), composed of $\alpha$ and $\beta$ subunits, is a heterodimer molecule, which is regulated by pVHL and plays a role in cellular response to hypoxia [1]. HIF was degraded by VCB-CUL2 complex mediated by ubiquitin. PVHL is involved in the formation of VCB-CUL2 complex and plays a tumor suppressor role by participating in the degradation of HIF. In an aerobic environment, VCB-CUL2 complex binds to the $\alpha$ subunit of HIF and degrades HIF through multiubiquitination markers. However, in the case of environmental hypoxia or loss of function of pVHL, the VCB-CUL2 complex cannot bind to the $\alpha$ subunit of HIF, resulting in the failure of HIF to be degraded, which leads to the continuous expression of pro-tumor molecules such as vascular endothelial growth factor, platelet-derived growth factor, erythropoietin, and transforming growth factor-a. Ultimately, the increase of tumor-promoting factors leads to accelerated proliferation, angiogenesis, and tumorigenesis [5].

**TABLE 2.4.1 The effects of VHL syndrome in each system.**

| Impact system | Major diseases | Images findings |
|---|---|---|
| Central nervous system | Central nervous system hemangioblastoma | Hemangioblastoma |
| Digestive system | Pancreatic cyst or tumor | Cyst or mass |
| Urinary system | Renal cyst or renal cell carcinoma | Single or multiple cyst or mass |
| Endocrine system | Pheochromocytoma | Adrenal mass |
| Reproductive system | Epididymal cystadenomas, broad ligament cystadenomas | Cystadenomas |
| Eyes | Retinal angiomas | Angiomas |
| Ears | Endolymphatic sac tumors | Mass in endolymphatic |

## 3. Central nervous system

Hemangioblastoma arising in the CNS is not only the most common tumor in VHL patients, but also the leading cause of morbidity and mortality in VHL patients. Hemangioblastoma affects 60%−80% of VHL patients, and is usually benign. However, because of the space-occupying effect of hemangioblastoma, it leads to clinical manifestations and even death in VHL patients. The prevalent sites of CNS hemangioblastoma are cerebellum (16%−69%), brainstem (5%−22%), spinal cord (13%−53%), cauda equina (11%), and supratentorial region (1%−7%). Tumor growth, cyst formation, and subsequent edema lead to clinical manifestations of CNS occupancy [5], with patients often experiencing headaches, vomiting, sensory or motor deficits, and ataxia. Hemangioblastomas in the CNS can cause headaches, vomiting, sensory or motor deficits, and ataxia.

Brain and spinal MRI are commonly used to monitor CNS hemangioblastoma [6]. MRI examination once a year is necessary for VHL patients over 10 years of age [7]. The recommended imaging modality for CNS hemangioblastoma is MRI with contrast, and lesions as large as 2 mm in size can also be detected and appear as enhancing lesions on T1-weighted images [5]. The typical manifestation is that about 80% of cystic lesions with enhanced mural nodules develop in the brain (Fig. 2.4.1), and 20% develop in the spinal cord [1]. Small and asymptomatic tumors need only careful observation. Symptomatic tumors or tumors located in important areas and large hemangioblastomas should be treated by surgical resection or with gamma knife ablation [7].

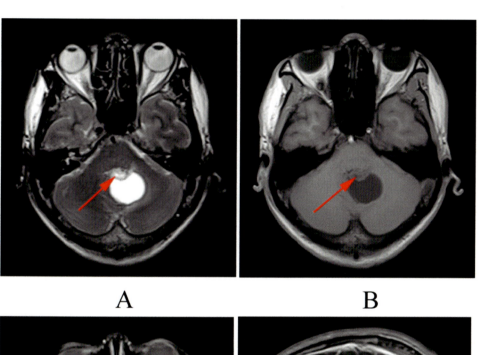

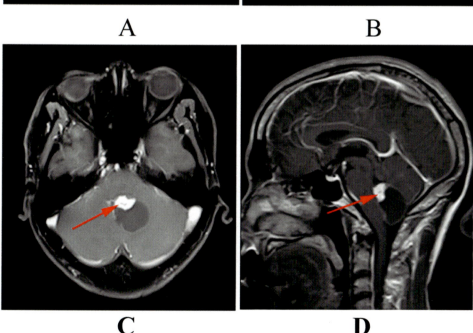

A        B

C        D

**FIGURE 2.4.1** (A and B) Axial cranial MRI scan shows a cyst with wall nodules in the right part of the cerebellum with T1WI low signal and T2WI high signal (*arrow*). (C and D) Axial postcontrast T1WI shows wall nodular enhancement of the cyst in the right part of the cerebellum, without enhancement of the cyst and cyst wall.

## 4. Digestive system

Pancreatic lesions occur in 17%—56% of VHL patient [1]. Pancreatic cysts are found at autopsy in 72% of VHL patients and they are considered the only sign in 12% of VHL patients [5,8,9]. Pancreatic cysts are predominantly multiple and usually asymptomatic. In addition to pancreatic cysts, pancreatic lesions in patients with VHL are also manifested as serous cystadenoma and neuroendocrine tumors. Patients with VHL rarely die from pancreatic neuroendocrine tumors (PNETs), but once PNET deteriorates or metastasizes, this will lead to poor prognosis, and 8% of VHL patients may experience malignancy and metastasis of the neuroendocrine tumor. Pancreatic cysts or pancreatic adenomas cause pancreatic parenchyma replacement, resulting in pancreatic exocrine or endocrine deficiency [5].

Pancreatic cysts are usually asymptomatic, so routine imaging examination is of great significance in the diagnosis of VHL patients. Ultrasound and CT can detect pancreatic cysts (Fig. 2.4.2), while CT can also detect pancreatic cysts and tumors [1]. PNET appears as enhancing mass on contrast-enhanced CT imaging. Once the mass is detected, further MRI examination can help to diagnose this lesion. For neuroendocrine tumors that cannot be detected by CT, the diagnosis is made with the help of FDG positron emission tomography [5].

Surgical treatment is generally not used for asymptomatic pancreatic cysts; however, patients with obstructive symptoms require surgical relief of compression. The choice of Whipple procedure or partial pancreatectomy to remove potentially metastatic PENTs is based on the size and location of the tumor. Tumor size greater than 3 cm, pathogenic mutations in exon 3, and tumors that double in growth rate within 500 days, all of these increase the risk of metastasis and meet the resection criteria [5].

## 5. Urinary system

Renal cysts and clear cell carcinoma can occur in the kidneys of VHL patients, with lesions ranging from simple cysts to completely solid lesions. Patients mostly present with abdominal pain or hematuria with a mass in the kidney area [10]. Renal cysts can be found in 59%—63% of patients with VHL [1]. Simple renal cysts usually have no signs or symptoms, but complex renal cysts can turn into solid kidney masses. Although kidneys often have multiple cysts, renal function is rarely affected, and chronic renal failure is not frequent [11]. RCC occurs in 25%—45% of VHL patients [1]. They are rarely the first symptom of VHL. The rate of VHL complicated by RCC is as high as 70% at the age of 60. For these patients, the main cause of death is RCC [8].

Renal cysts are usually multiple lesions of varying sizes in both kidneys, and ultrasound, CT, and MRI can all be used to detect renal cysts (Fig. 2.4.3). The preferred imaging tool for RCC is CT. RCC usually occurs on both sides, and can manifest as multiple lesions, both solid or cystic [1]. Since RCC is asymptomatic for a significant period of time, serial kidney imaging examination is significant in monitoring for the development of RCC. Simple renal cysts in VHL patients generally do not require treatment, but complex cysts need regular monitoring because they may contain solid components of RCC. Diagnosing kidney disease before clinical symptoms appear is likely to improve the prognosis. The recommended method for diagnosing kidney lesions in VHL patients is abdominal CT imaging, which quantifies the size and number of tumors or cysts. Differentiation between simple and complex cysts relies on CT and MRI. In terms of histology, both RCC

**FIGURE 2.4.2** (A) Axial CT shows two hypodense lesions in the body of the pancreas (*arrow*); (B) Postcontrast axial CT reveals two hypodense lesions in the body of the pancreas without enhancement (*arrow*).

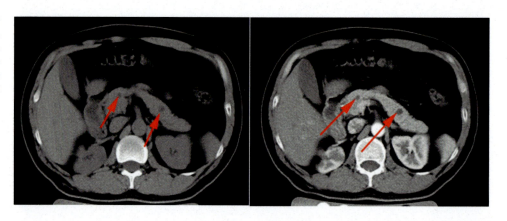

A                    B

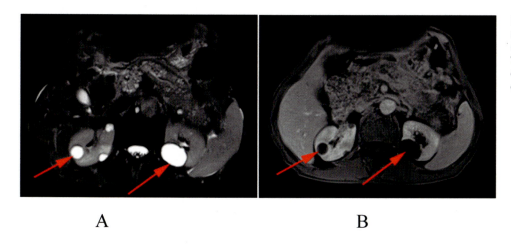

FIGURE 2.4.3 (A) Axial T1WI shows multiple cysts in both kidneys (*arrow*); (B) Postcontrast axial T1WI reveals shows the multiple cysts in both kidneys without enhancement (*arrow*).

and renal cysts in VHL patients are clear cell type. VHL-associated RCC and sporadic clear cell carcinoma are also histologically similar in that they are usually of low-grade histology [5,8,12].

Tumors smaller than 3 cm do not require intervention, while tumors larger than 3 cm can be treated by partial nephrectomy. This reduces the risk of metastasis while preserving renal function. For patients with VHL who underwent kidney-preserving surgery for lesions larger than 3 cm, the 10-year survival rate reached 81% under the condition of preserving renal function. For tumors smaller than 3 cm, percutaneous radiofrequency ablation or laparoscopic radiofrequency ablation has been shown to be effective and reduce complication rates. However, RF ablation of lesions requires frequent monitoring and intervention [5].

## 6. Endocrine system

Pheochromocytomas occur in 0%−60% [1]. They can occur bilaterally and sometimes present as multifocal. Pheochromocytomas in VHL patients tend to appear in around 20 years of age and have a low chance of transforming into malignant tumors. Pheochromocytomas affecting the adrenal glands may be asymptomatic or may present with symptoms such as high blood pressure, tachycardia, headache, sweating, pale face, and nausea due to excessive production of catecholamines by the tumor [5,9].

Both laboratory tests and imaging examinations can assist in the diagnosis of pheochromocytoma. Laboratory studies, including serum and urine catecholamines, can help diagnose pheochromocytoma, which is especially important when imaging examinations does not reveal adrenal lesions. The best way to detect pheochromocytoma is plasma-free adrenaline, with a sensitivity of 97%, which is more sensitive than catecholamine in 24 hours urine because the latter may produce false negative results [1,5]. Pheochromocytoma can be identified by CT or MRI [1]. Both adrenal pheochromocytoma and extraadrenal pheochromocytoma can be diagnosed with enhanced abdominal CT (Fig. 2.4.4), but a more

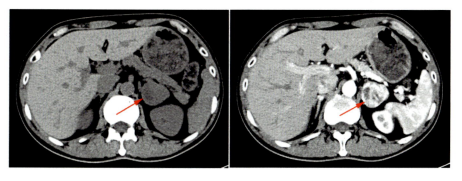

FIGURE 2.4.4 (A) Axial CT shows a mixed density mass in the left adrenal gland (*arrow*); (B) Postcontrast axial CT reveals significant inhomogeneous enhancement of the left adrenal mass (*arrow*).

sensitive modality is abdominal-enhanced MRI, which has a sensitivity of between 90% and 100%. Meta-iodobenzylguanidine scintigraphy is more useful for the identification of extraadrenal tumors [5].

Laparoscopic partial adrenalectomy is the preferred treatment for pheochromocytoma. To prevent serious catecholamine-mediated complications such as arrhythmias, hypertensive crisis, and myocardial infarction, the combination of $\alpha$-adrenergic and $\beta$-adrenergic blockade therapy can be used in the perioperative period. Pharmacological intervention is required 10−14 days prior to surgery. Pheochromocytoma in VHL patients undergoing early cortical preservation with partial adrenalectomy has a low recurrence rate [5,13].

## 7. Reproductive system

25%−60% of male VHL patients develop unilateral or bilateral epididymal cystadenoma during adolescence. This kind of tumor is usually benign, asymptomatic, and is detected by chance, so surgery is usually not required. Approximately 17% of patients often cannot be detected on physical examination because of the small size of the lesion. Ultrasound is preferred for localization of these lesions. Morphologically, the tumor is similar to other VHL tumors [8,10,11]. Papillary cystadenoma, which is detected in unilateral or bilateral fallopian tube mesangial or broad ligaments in female patients with VHL, is rare and its incidence is unknown. The main manifestation of the tumor is a pelvic mass with abdominal discomfort or pain in the adnexal region [8,11,12].

Physical examination of epididymal cystadenoma is usually not positive due to its small size and difficulty of palpation. Therefore, ultrasonography is the preferred screening modality. Grossly, epididymal cystadenomas consist of solid components and cystic components composed of colloid material, but the solid components are predominant. Papillary cystadenoma of the broad ligament has a prominent papillary structure on histology. The center of the lesion is predominantly hyaline interstitial and fibrovascular. Imaging methods for this disease are mainly abdominal MRI or pelvic ultrasound. Epididymal cystadenomas and broad ligament cystadenomas are usually benign and do not require specific treatment. Ultrasound or CT is usually used to monitor the lesions clinically [5].

## 8. Eyes

Retinal capillary hemangioma, occurring in 45%−59% of patients [1], is the earliest manifestation of VHL syndrome [14]. Retinal capillary hemangioma can lead to serum leakage, fibrous glial band formation, retinal detachment and vitreous hemorrhage, and eventually patients develop glaucoma or permanent vision loss [1]. Complications such as blindness or severe visual impairment occur in 5%−8% VHL patients. Early onset of retinal capillary hemangioma, bilateral retinal involvement, and missense mutations in VHL can lead to poorer vision [5]. Retinal hemangioblastoma, retinal detachment, macular edema, or cataract can be detected by ophthalmoscope. Diagnosis of glaucoma can be made by intraocular pressure measurement [1]. Retinal screening is recommended at least once a year. Retinopathy related to arterial hypertension during the development of pheochromocytoma is another reason for visual impairment [7].

## 9. Ears

Endolymphatic system is closely related to the generation and absorption of endolymph in cochlea and semicircular canal. 6%−15% of VHL patients have endolymphatic sac tumor (ELST), but this tumor is rare in other populations and originates from endolymphatic epithelium in vestibular aqueduct. They are benign tumors that may be locally aggressive, but do not metastasize [5]. ELST occurs in the inner ear, and patients may experience clinical manifestations such as tinnitus, vertigo, or hearing loss. The size of ELST more than 3 cm can cause facial nerve abnormalities, leading to facial paralysis. It is necessary to conduct hearing assessment when hearing loss is caused by ELST [1].

CT and MRI are essential for the diagnosis of ELST [8]. The tumors occur primarily in the endolymphatic tract, but they are richly vascularized and often erode the nearby temporal bone. When ELST is small, the tumor is completely located in the endolymphatic sac. However, when it is large, it will grow outward and erode the temporal bone [5]. On contrast-enhanced CT, the enhancement degree of ELST is similar to that of brain parenchyma; however, ELST may show areas with stronger or weaker enhancement than brain parenchyma. CT may also show large tumors in the center of the endolymphatic duct expanding toward the temporal bone [12]. For patients with ELST associated with clinical symptoms, thin-slice MRI through the internal auditory canal is recommended [1]. On precontrast T1-weighted MRI, intensity levels may be homogeneous or inhomogeneous. Postcontrast T1-weighted MRI will show a heterogeneous enhancement pattern or homogeneity [5].

# References

[1] Mikhail M, ISingh AK. von Hippel Lindau syndrome. StatPearls Publishing Copyright © 2021. StatPearls Publishing LLC; 2021.

[2] Perlman S. von Hippel-Lindau disease and Sturge—Weber syndrome. Handb Clin Neurol 2018;148:823—6.

[3] Binderup ML, Galanakis M, Budtz-Jørgensen E, et al. Prevalence, birth incidence, and penetrance of von Hippel-Lindau disease (vHL) in Denmark. Eur J Hum Genet 2017;25(3):301—7.

[4] Qiu J, Zhang K, Ma K, et al. The genotype-phenotype association of von Hipple Lindau disease based on mutation locations: a retrospective study of 577 cases in a Chinese population. Front Genet 2020;11:532588.

[5] Varshney N, Kebede AA, Owusu-Dapaah H, et al. A review of von Hippel-Lindau syndrome. J Kidney Cancer VHL 2017;4(3):20—9.

[6] Rednam SP, Erez A, Druker H, et al. von Hippel-Lindau and hereditary pheochromocytoma/paraganglioma syndromes: clinical features, genetics, and surveillance recommendations in childhood. Clin Cancer Res 2017;23(12):e68—75.

[7] Ben-Skowronek I, Kozaczuk S. von Hippel-Lindau syndrome. Horm Res Paediatr 2015;84(3):145—52.

[8] Chittiboina P, Lonser RR. von Hippel-Lindau disease. Handb Clin Neurol 2015;132:139—56.

[9] Cassol C, Mete O. Endocrine manifestations of von Hippel-Lindau disease. Arch Pathol Lab Med 2015;139(2):263—8.

[10] Maher ER, Neumann HP, Richard S. von Hippel-Lindau disease: a clinical and scientific review. Eur J Hum Genet 2011;19(6):617—23.

[11] Shanbhogue KP, Hoch M, Fatterpaker G, et al. von Hippel-Lindau disease: review of genetics and imaging. Radiol Clin N Am 2016;54(3):409—22.

[12] Lonser RR, Glenn GM, Walther M, et al. von Hippel-Lindau disease. Lancet 2003;361(9374):2059—67.

[13] Benhammou JN, Boris RS, Pacak K, et al. Functional and oncologic outcomes of partial adrenalectomy for pheochromocytoma in patients with von Hippel-Lindau syndrome after at least 5 years of followup. J Urol 2010;184(5):1855—9.

[14] Jain K, Singh MD. von Hipple-Lindau: unusual case presentation with peripheral and juxtapapillary retinal hemangioma. Oman J Ophthalmol 2018;11(2):166—8.

# Chapter 3

# Muscular system

Daniel Phung, Gordon Crews, Raymond Huang and Nasim Sheikh-Bahaei

*Keck School of Medicine of USC, Los Angeles, CA, United States*

## 1. Poliomyelitis

Polio is a debilitating viral disease with the most severe form, poliomyelitis, characterized by acute-onset limb paralysis [1]. Poliovirus is an enterovirus spread exclusively by humans via the fecal—oral route and infects lymphatic tissues, initially in the oropharynx and subsequently throughout the intestinal tract. The virus is thought to then spread along afferent nerves to the central nervous system including the anterior horn cells of the spinal cord with resultant damage and cell death. Infected patients have a wide range of presentations from asymptomatic (95%) to demonstrating mild respiratory or gastrointestinal symptoms (5%) to acute meningitis and paralysis (1% or less). Lower extremity paralysis occurs more commonly than upper extremity paralysis, and in rare cases, infection may involve the brainstem leading to bulbar paralysis. Death is exceedingly rare and is due to respiratory paralysis and failure [1—4]. Postpolio syndrome is a delayed manifestation of disease occurring 2-3 decades after infection which is characterized by progressive muscle atrophy, presumably related to ongoing neuronal degeneration [5]. The disease has been nearly eradicated throughout the world due to the spectacular success of global vaccination programs; however, the disease remains endemic in some areas of Africa and the middle East where poor sanitation exists [1,2]. Diagnosis is typically made by polymerase chain reaction of serial stool samples. Management of disease focuses on rigorous physical therapy and specific orthopedic surgical interventions dependent upon the extent of paralysis.

Imaging in the acute illness often demonstrates no significant abnormality. The chronic sequelae of poliomyelitis, however, are evidenced by striking lipoatrophy or "vanishing" of various muscle groups. MRI can demonstrate T2/FLAIR hyperintensity of the spinal cord ventral gray matter and precentral gyrus [2,3,5]. A current differential consideration in the pediatric population is acute flaccid myelitis related to enterovirus D68 infection. This entity also presents with acute onset of lower extremity predominant paralysis with viral prodrome. These patients may also present with spinal cord ventral gray matter T2 hyperintensity to further complicate matters. Clinical history and isolation of the virus in cerebrospinal fluid or stool are essential for discrimination [6,7] (Fig. 3.1).

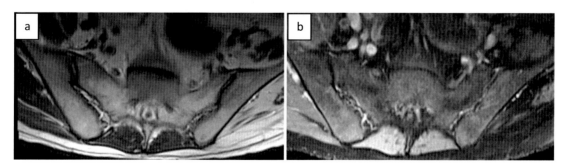

**FIGURE 3.1** (A) left (axial T1) and b right (axial T1 fat-suppressed postcontrast). Patient with history of poliomyelitis demonstrating marked asymmetric lipoatrophy of the left iliacus and gluteal muscles.

**Multi-system Imaging Spectrum associated with Neurologic Diseases.** https://doi.org/10.1016/B978-0-323-91795-7.00014-2

## 2. Scleroderma

Scleroderma describes a family of autoimmune disorders characterized by inflammation, vasculitis, and fibrosis involving the skin with female predominance [8]. Scleroderma is divided into localized and systemic forms based upon the coexistence of extracutaneous manifestations such as visceral organ fibrosis. However, central, peripheral, and autonomic nervous system involvement has been increasingly recognized in both forms with some phenotypic variance. Localized scleroderma has itself additional subsets including linear scleroderma "en coup de sabre" and progressive hemifacial atrophy (also known as Parry-Romberg syndrome) [9,10]. Treatment is tailored toward symptom management and typically involves immunosuppressant therapy, most often corticosteroids. Antiepileptics are also often used for management of seizures. Surgery may be performed for entrapment neuropathies [8,11,12].

Patients with localized scleroderma present in childhood with oval cutaneous fibrotic plaques with truncal and proximal extremity predominance. In the acute phase, a "violaceous ring" may be identified surrounding the plaques. Patients with linear scleroderma "en coup de sabre" present with one or more unilateral linear fibrotic plaques involving the fronto-parietal scalp which have been likened to a cut from a saber. These linear plaques are associated with alopecia and thinning of the subjacent calvarium. Progressive hemifacial atrophy is often associated with linear scleroderma and describes a cutaneous and muscular facial atrophy which may progress for several years and is often unilateral in distribution [9–11]. These cutaneous manifestations in localized scleroderma precede neurologic manifestations, the most common of which are epilepsy, headache, and cranial neuropathies. However, hemiplegia and cerebellar ataxias have also been described [11]. Neuroimaging findings and cutaneous findings are often ipsilateral. CT may demonstrate central gray matter calcifications and cortical depression. MRI often demonstrates T2/FLAIR hyperintense white matter lesions (presumably related to chronic perivascular inflammation), gray–white matter blurring, and cerebral atrophy [9,13].

Systemic scleroderma patients often present in early to middle adulthood with heterogeneous symptoms depending upon the extent of cutaneous, vascular, and visceral organ involvement. Patients may present with aforementioned fibrotic plaques, Raynaud phenomenon, and varying cardiopulmonary, renal, and gastrointestinal dysfunction. Patients are at significantly increased risk of depression and anxiety with prevalence of approximately 30%–40% and 25%–64% respectively. Neurologic manifestations include headache, cognitive impairment, trigeminal neuropathy, and rarely transverse myelitis. Autonomic dysfunction is common, occurring in up to 79% of patients, and is characterized by parasympathetic hypoactivity and sympathetic hyperactivity. Peripheral entrapment neuropathies such as carpal tunnel syndrome are also common. Neuroimaging findings in systemic scleroderma are similar to those seen in localized scleroderma [10,13] (Fig. 3.2).

## 3. Rhabdomyosarcoma

Rhabdomyosarcoma (RMS) is the most common soft-tissue sarcoma in children with a prevalence of 4.5 cases per million in patients under 20 years and accounting for 7% of all childhood malignancies. RMS arises from skeletal myoblasts by two pathways, resulting in different subtypes. Translocation mutations that generate an oncogenic fusion protein, PAX03-FOX01 or PAX07-FOX01, in the so-called "fusion-positive RMS" typically correspond to the alveolar subtype of RMS. Alternatively, a combination of activating mutations on oncogenes such as FGFR4 and loss-of-function mutations in tumor suppressor genes such as TP53 in so-called "fusion-negative RMS" typically correspond to the embryonic subtype of RMS. Embryonic RMS demonstrates a predilection for head and neck involvement, whereas alveolar RMS tends to involve the extremities [14–17]. Patients often present with painless soft-tissue mass. In the head and neck, RMS may compress nerves and result in neurologic symptoms. In patients with orbital disease, strabismus and ophthalmoplegia in addition to exophthalmos can be seen [14,17–19].

Diagnosis requires histologic analysis [14–16,20]. On MRI, these masses appear T1 isointense to hyperintense with heterogeneous T2 hyperintensity and enhancement due to internal hemorrhage and necrosis [17,21,22]. Patients with

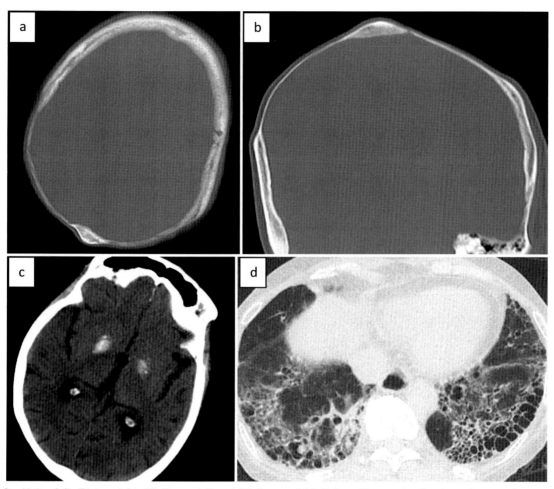

**FIGURE 3.2**   A) (*top left*), (B) (*top right*), (C) (*bottom left*), and (D) (*bottom right*). Patient with systemic scleroderma demonstrating severe biparietal calvarial thinning (Fig. 3.2A and B). Additional manifestations are seen in bilateral basal ganglia calcifications (Fig. 3.2C) and pulmonary fibrosis (Fig. 3.2D).

localized RMS undergo a combination of surgical excision, chemotherapy, and radiation with a 5-year survival rate of 70% −80%. Patients with metastatic disease fare much worse with a 5-year survival rate of 30%. Approximately 20% of patients are found with metastatic disease at the time of diagnosis, and patients with initially local disease may develop metastases over the course of treatment [15−17,20] (Fig. 3.3).

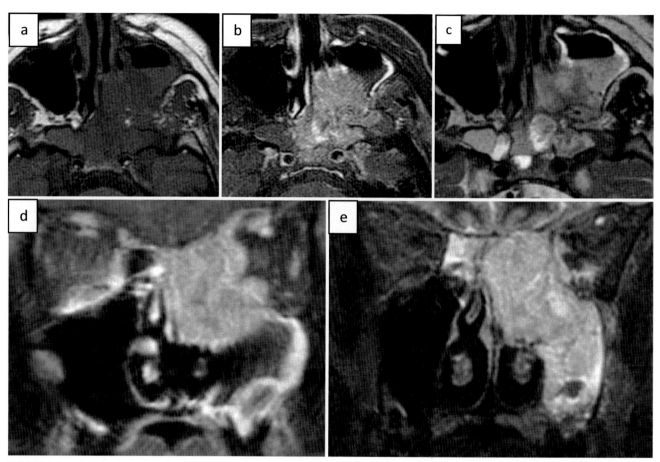

**FIGURE 3.3** A) top left (axial T1), b top middle (axial T1 fat-suppressed postcontrast), c top right (axial T2), d bottom left (coronal T1 fat-suppressed postcontrast), e bottom right (coronal T2 fat suppressed). Patient with left nasal cavity rhabdomyosarcoma which demonstrates osseous erosion of the left medial orbital wall, nasal septum, and left maxillary sinus medial and lateral walls. The mass demonstrates avid enhancement and heterogeneous T2 hyperintensity with invasion of the pterygomaxillary fissure anteriorly and basisphenoid posteriorly (Fig. 3.3A—C). Coronal images demonstrate replacement of the superior and middle turbinates by tumor as well as erosion of the left medial orbital wall at the lateral aspect of the mass and left fovea ethmoidalis at the superior aspect. In keeping with the aggressive nature of the mass, the coronal images also demonstrate extension of enhancing tissue into the left medial orbit and suggestion of epidural extension past the left fovea ethmoidalis (Fig. 3.3D—E).

# References

[1] Montalvo M, Cho TA. Infectious myelopathies. Neurol Clin 2018;36(4):789—808.

[2] Bao J, Thorley B, Isaacs D, et al. Polio - the old foe and new challenges: an update for clinicians. J Paediatr Child Health 2020;56(10):1527—32.

[3] Kidd D, Williams AJ, Howard RS. Poliomyelitis. Postgrad Med 1996;72(853):641—7.

[4] Mehndiratta MM, Mehndiratta P, Pande R. Poliomyelitis: historical facts, epidemiology, and current challenges in eradication. Neurohospitalist 2014;4(4):223—9.

[5] Lo JK, Robinson LR. Post-polio syndrome and the late effects of poliomyelitis. Muscle Nerve 2018;58(6):751—69.

[6] Helfferich J, Knoester M, Van Leer-Buter CC, et al. Acute flaccid myelitis and enterovirus D68: lessons from the past and present. Eur J Pediatr 2019;178(9):1305—15.

[7] Murphy OC, Messacar K, Benson L, et al. Acute flaccid myelitis: cause, diagnosis, and management. Lancet 2021;397(10271):334—46.

[8] Fett N. Scleroderma: nomenclature, etiology, pathogenesis, prognosis, and treatments: facts and controversies. Clin Dermatol 2013;31(4):432—7.

[9] Duman IE, Ekinci G. Neuroimaging and clinical findings in a case of linear scleroderma en coup de sabre. Radiol Case Rep 2018;13(3):545—8.

[10] Amaral TN, Peres FA, Lapa AT, Marques-Neto JF, Appenzeller S. Neurologic involvement in scleroderma: a systematic review. Semin Arthritis Rheum 2013;43(3):335—47.

[11] Kister I, Inglese M, Laxer RM, Herbert J. Neurologic manifestations of localized scleroderma: a case report and literature review. Neurology 2008;71(19):1538—45.

[12] Zhao M, Wu J, Wu H, Sawalha AH, Lu Q. Clinical treatment options in scleroderma: recommendations and comprehensive review. Clin Rev Allergy Immunol 2022;62(2):273—91.

[13] Khodarahmi I, Alizai H, Chalian M, et al. Imaging spectrum of calvarial abnormalities. Radiographics 2021;41(4):1144—63.

[14] Reilly BK, Kim A, Peña MT, et al. Rhabdomyosarcoma of the head and neck in children: review and update. Int J Pediatr Otorhinolaryngol 2015;79(9):1477–83.

[15] Skapek SX, Ferrari A, Gupta AA, et al. Rhabdomyosarcoma. Nat Rev Dis Prim 2019;5(1):1.

[16] Yechieli RL, Mandeville HC, Hiniker SM, et al. Rhabdomyosarcoma. Pediatr Blood Cancer 2021;68(Suppl. 2):e28254.

[17] Jawad N, McHugh K. The clinical and radiologic features of paediatric rhabdomyosarcoma. Pediatr Radiol 2019;49(11):1516–23.

[18] Larson JH, Rutledge R, Hunnell L, Choi DK, Kellogg RG, Naran S. Nasal embryonal rhabdomyosarcoma in the pediatric population: literature review and report of midline presentation. Plast Reconstr Surg Glob Open 2021;9(4):e3534.

[19] Leviton A, Davidson R, Gilles F. Neurologic manifestations of embryonal rhabdomyosarcoma of the middle ear cleft. J Pediatr 1972;80(4):596–602.

[20] Rogers TN, Dasgupta R. Management of rhabdomyosarcoma in pediatric patients. Surg Oncol Clin 2021;30(2):339–53.

[21] Joseph AK, Guerin JB, Eckel LJ, et al. Imaging findings of pediatric orbital masses and tumor mimics. Radiographics 2022:210116.

[22] Inarejos Clemente EJ, Navallas M, Barber Martínez de la Torre I, et al. MRI of rhabdomyosarcoma and other soft-tissue sarcomas in children. Radiographics 2020;40(3):791–814.

# Chapter 4

# Respiratory system

**Tian-le Wang**

*Affiliated Hospital 2 of Nantong University, Nantong, Jiangsu, China*

## Chapter 4.1

## Aspergillosis

*Aspergillus* is a conditional pathogen, often occurring in patients with severe immunosuppression and long-term neutropenia, such as human immunodeficiency virus infection, organ transplantation, alcoholism, and hematologic malignancies. It may also be seen in patients without immunosuppression but with chronic disease, and it can develop in people with normal immune function. It can involve the skin, mucous membranes, eyes, nose, bronchial tubes, lungs, and nervous system. In recent years, more and more cases of immunocompetent *Aspergillus* infections have been reported, with risk factors such as hormonal therapy and antitubercular drugs [1]. It may run a chronic or insidious course, usually presenting as granulomatous masses or meningitis [2,3]. There might be delayed diagnosis or missed diagnosis due to clinical and atypical imaging [4]. Most immunocompetent cases of *Aspergillus* infection in the brain have sinus infection [5,6], while isolated foci in brain parenchyma due to nonsinus or pulmonary spread of infection were also reported [7]. The main symptoms of aspergillosis infection are related to the site of infection, and the most common symptoms are fever, cough, green pus sputum, and wheezing. More severe disease may also cause hemoptysis, chest pain, skin erythema, headache, and coma. Common complications include bronchiectasis and sepsis, which may cause death if left untreated.

Imaging findings:

(1) Intracranial Aspergillosis: Brain imaging can be used as an aid in the diagnosis of *Aspergillus* infection and help localize the lesions in the brain, but it cannot be used for definitive diagnosis [8]. Brain MRI findings are highly variable, which may show ring-enhancing lesions or abscess formation, meningitis and or meningoencephalitis, and small infarcts with or without hemorrhage due to vasculitis [5]. Brain parenchymal lesions may show low signal on T1-weighted imaging (T1WI) and high signal on T2-weighted imaging on MRI due to local edema (Fig. 4.1.1A). In contrast, intracerebral abscesses caused by bacterial infections are characterized by significant diffusion restriction in the center of the abscesses. There is significant enhancement, which may be nodular, jagged, or lace-like (Fig. 4.1.1B) [9,10]. CT has no significant advantage in diagnosing intracranial aspergillosis, but it is helpful in detecting hemorrhage, calcification, and bone destruction.

(2) Pulmonary invasive aspergillosis: The typical CT presentation is a "halo sign" manifested by a ground glass-like halo around the nodule or a wide basal wedge-shaped solid change in the pleura (Fig. 4.1.1C). The "halo sign" manifests as *Aspergillus*-infected pulmonary nodules with infarction and coagulative necrosis surrounded by alveolar hemorrhage. Eventually, an "air bubble" sign may appear as the central necrotic tissue may be separated from the surrounding lung parenchyma, forming an air crescent sign. The "halo sign" has been recognized as an early sign of invasive aspergillosis. When the CT "halo sign" is found in immunocompromised patients, it is highly suggestive of invasive aspergillosis.

**Multi-system Imaging Spectrum associated with Neurologic Diseases. https://doi.org/10.1016/B978-0-323-91795-7.00012-9**

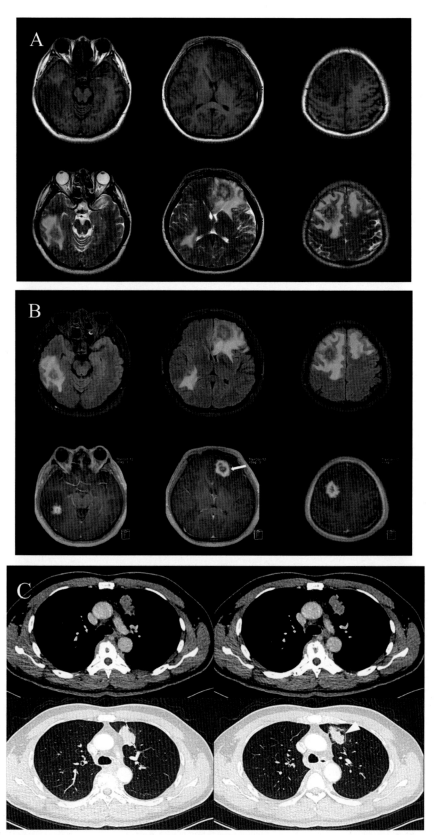

**FIGURE 4.1.1**    A 45-year-old man with an *aspergillus* infection. (A) Brain MRI showing multiple brain lesions with associated edema. (B) Brain MRI showing multiple brain lesions with the zigzag and lace-like enhancing pattern (*white arrow*). (C) Chest CT shows the "halo sign" (*arrow*) with vacuoles around the anterior segment nodule in the left upper lobe.

# References

[1] Panda PK, Mavidi SK, Wig N, et al. Intracranial aspergillosis in an immunocompetent young woman. Mycopathologia 2017;182(5−6):527−38.

[2] Azarpira N, Esfandiari M, Bagheri MH, et al. Cerebral aspergillosis presenting as a mass lesion. Braz J Infect Dis 2008;12(4):349−51.

[3] Jain KK, Mittal SK, Kumar S, et al. Imaging features of central nervous system fungal infections. Neurol India 2007;55(3):241−50.

[4] Shamim MS, Enam SA, Ali R, et al. Craniocerebral aspergillosis: a review of advances in diagnosis and management. J Pakistan Med Assoc 2010;60(7):573−9.

[5] Marzolf G, Sabou M, Lannes B, et al. Magnetic resonance imaging of cerebral aspergillosis: imaging and pathological correlations. PLoS One 2016;11(4):e0152475.

[6] Siddiqui AA, Bashir SH, Shah AA, et al. Diagnostic MR imaging features of craniocerebral Aspergillosis of sino-nasal origin in immunocompetent patients. Acta Neurochir 2006;148(2):155−66.

[7] Bokhari R, Baeesa S, A-l Maghrabi J, et al. Isolated cerebral aspergillosis in immunocompetent patients. World Neurosurg 2014;82(1−2):e325−33.

[8] Gavito-Higuera J, Mullins CB, Ramos-Duran L, et al. Fungal infections of the central nervous system: a pictorial review. J Clin Imaging Sci 2016;6:24.

[9] Gärtner F, Forstenpointner J, Ertl-Wagner B, et al. CT and MRI findings in cerebral aspergilloma. Fortschr Röntgenstr 2018;190(10):967−70.

[10] Kickingereder P, Sahm F, Wiestler B, et al. Evaluation of microvascular permeability with dynamic contrast-enhanced MRI for the differentiation of primary CNS lymphoma and glioblastoma: radiologic-pathologic correlation. AJNR Am J Neuroradiol 2014;35(8):1503−8.

Chapter 4.2

# Langerhans cell histiocytosis

Langerhans cell histiocytosis (LCH) is a histiocytic disease, formerly known as histiocytosis X, which includes eosinophilic granulomatosis, Hand-Schuller-Christian disease, and Letterer—Siwe disease. It is characterized by clonal proliferation of histiocytes and excessive accumulation of pathological Langerhans cells, and is classified in the 2017 edition of the WHO classification of histiocytic diseases and macrophage—dendritic cell lineage tumors as a group L together with Erdheim—Chester disease [1—3]. LCH is currently considered to be an inflammatory myeloid tumor. LCH can be seen at any age, but mostly occurs in infancy and childhood, with a peak age of diagnosis at 1—3 years and rarely in adults. The etiology of LCH is still unclear, but it is generally believed to be related to immune dysfunction, and the diagnosis requires clinical, imaging, and pathologic involvement of three systems: single-system involvement (e.g., isolated skin or bone disease) or multisystem involvement, which may manifest as simultaneous involvement of two or more systems: skeletal, digestive (liver, spleen), pulmonary, bone marrow, endocrine, central nervous system (CNS), skin, and lymph nodes [2]. The most frequently involved systems are bone (about 80%), skin (about 33%), and pituitary gland (about 25%) [4]. LCH is polymorphic in its manifestations in all systems, but it is characterized by a monoclonal cluster of CD1a (+) mononuclear cells [1]. The diagnosis of LCH is based on histological and immunophenotypic test of the tissue, which is characterized by the morphology of Langerhans cells and CD1a (+), Langerin granules [1].

Intracranial granulomatous infiltration of LCH can cause lesions in the brain. When LCH involves the CNS, the lesions are divided into primary and secondary in origin, and mainly intraaxial and extraaxial in location. Extraaxial tissues without the blood—brain barrier are the most easily invaded areas by LCH, such as pituitary gland, meninges, pineal gland, choroid plexus, etc., while hypothalamic pituitary gland is the most frequently involved site and the earliest site. The characteristic manifestation is central uveitis, which is also an indicator of disease activity. The second symptom is abnormal anterior pituitary hormone secretion, such as growth hormone deficiency, secondary hypothyroidism, and reduced sex hormone secretion. Normally, the posterior pituitary gland shows hyperintense on T1WI, which is associated with the presence of antidiuretic hormone granules. When LCH is present, there is a loss of high signal on T1WI in the posterior pituitary lobe and there is also thickening of the pituitary stalk (>3 mm) with enhancement (Fig. 4.2.1). When the lesion is large, it may also show linear stenosis of the infundibulum and a mass at infundibulum or hypothalamus and with significant enhancement, which needs to be differentiated from other lesions in suprasellar region such as germ cell tumors, craniopharyngioma, meningioma, and nodular disease. LCH involvement of dura and choroid plexus shows localized masses, all of which are formed by pathological yellow granulomas. Dural masses are most often found on cerebral convexity, falx and tentorium, located either subdural or epidural, and may be solitary or multiple. Choroid plexus lesions are mainly located in lateral ventricular triangle, with some reports of involvement of the lateral ventricular wall, and lesions that block foramen magnum may cause hydrocephalus [5]. LCH may show an increase in volume (>10 mm) or cystic changes when pineal gland is involved, and pathologically solid lesions appear as yellow granulomas with significant enhancement [6] (Fig. 4.2.1). Intraaxially, gray matter is more commonly involved by LCH, mainly cerebellar dentate nucleus, basal ganglia, and pons, with a bilateral symmetrical distribution. The imaging presentation is equal or slightly hypointense on T1WI and hypointense on T2WI as compared to gray matter, and early lesions are mildly enhancing with clear margins.

LCH pulmonary lesions account for approximately 10% of cases and are more common in childhood than in infancy [7,8]. The lesions can be limited but are more often part of a systemic lesion. High-resolution lung CT is an important tool for the diagnosis of pulmonary involvement by LCH. Pulmonary CT shows a variety of manifestations, which vary with the degree of disease progression and typically present as pulmonary nodules or vesicles. During early stage, the lesions mostly appear as small solid nodules or dot-like dense shadows. The nodules are diffuse, bilaterally symmetric, often with blurred margins, and may be disseminated or numerous, ranging from 0.5 to 10 mm (Fig. 4.2.1). The vesicles may coexist with nodules or exist alone, with a diameter of less than 10 mm or fused to 20—30 mm, and may be round or oval or irregular in shape, thin-walled or thick-walled or with uneven wall thickness. During late stage, pulmonary fibrosis may develop with the progress of time, eventually leading to rough striated shadow or honeycomb changes. The presence of new nodules on top of late vesicles suggests disease progression.

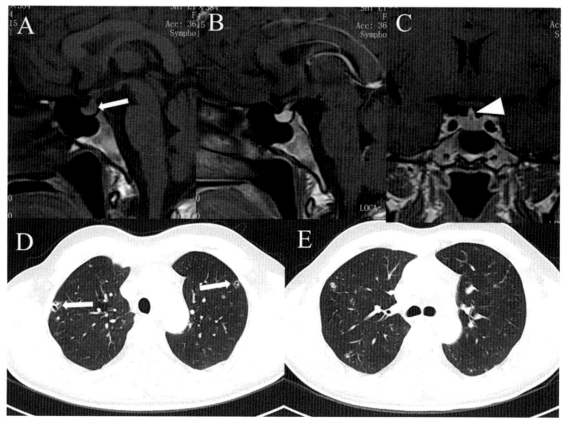

**FIGURE 4.2.1** A 16-year-old male patient with langerhans cell histiocytosis (LCHs) involving the brain and lung. Pituitary MRI (A−C): Absence of hyperintensity in part of posterior pituitary on T1WI (*white arrow*), pituitary stalk thickening with enhancement (*arrow*). Chest CT (D, E): Multiple vesicles and nodules in both lungs, less than 10 mm in diameter (*white arrow*).

# References

[1] Jain A, Kumar S, Aggarwal P, et al. Langerhans cell histiocytosis: an enigmatic disease. South Asian J Cancer 2019;8(3):183−5.

[2] Grana N. Langerhans cell histiocytosis. Cancer Control 2014;21(4):328−34.

[3] Howarth DM, Gilchrist GS, Mullan BP, et al. Langerhans cell histiocytosis: diagnosis, natural history, management, and outcome. Cancer 1999;85(10):2278−90.

[4] Donadieu J, Chalard F, Jeziorski E. Medical management of langerhans cell histiocytosis from diagnosis to treatment. Expet Opin Pharmacother 2012;13(9):1309−22.

[5] Zeng H, Gan Y, Huang W, et al. MRI features of Langerhans cell histiocytosis in central nervous system in children Chinese. J Radiol 2016;50(4):252−5.

[6] Prayer D, Grois N, Prosch H, et al. MR imaging presentation of intracranial disease associated with langerhans cell histiocytosis. AJNR 2004;25(5):880−91.

[7] Bano SA, Chaudhary V, Narula MK, et al. Pulmonary Langerhans cell histiocytosis in children: a spectrum of radiologic findings. Eur J Radiol 2014;83(1):47−56.

[8] Schmidt S, Eich G, Geoffray A, et al. Extraosseous langerhans cell histiocytosis in children. Radiographics 2008;28(3):707−26.

Chapter 4.3

# Metastatic tumor

Brain metastases (BMs) are the most common intracranial tumors in adults and spread to the CNS via the hematogenous route [1]. The incidence of BM varies by primary tumor. About 20% patients with cancer develop BMs. Most BMs occur in patients with lung cancer, breast cancer, colorectal cancer, melanoma, and renal cell carcinomas [2]. BM occurs most commonly from lung cancer, breast cancer, and melanoma [3]. BMs are present in 10%−20% of patients with nonsmall cell lung cancer (NSCLC) upon initial presentation, and the incidence is up to 50% in terminal NSCLC [4]. About 20% −65% of lung cancer patients will develop BMs during course of disease, which is the most common type of cancer developing BMs [5−7].

Lung tissue is rich in blood supply and lymph node. Once lung cancer cells invade adjacent venules, capillary, or lymphatic vessels and form tumor thrombus, they can easily reach distant organs through blood circulation. Lung adenocarcinoma and small cell lung cancer are the most common types for developing BM. Due to the invasive growth of squamous cell carcinoma, distant metastasis is relatively rare. However, cancers in other parts must enter the pulmonary circulation through the veins. After the filtering effect of the pulmonary capillary network, most of the tumor emboli stay in the lungs to be dissolved. Lung cancer cells invade the circulation through the left atrium and disperse to various parts of the body. The blood supply of the brain is large, accounting for 1/6 to 1/4 of the total blood circulation of the whole body; thus, the tumor thrombus may stay in the terminal branches of the main cerebral arteries, especially in the terminal branches of the middle cerebral artery. Therefore, the most common locations of BMs from lung cancer are in the frontal lobe, parietal lobe, and temporal lobe. Most supratentorial metastases occur at the junction of gray and white matter. Tumor glycoproteins in the lung tissue stimulate the growth of metastatic cancer cells, and may reach the brain with the blood circulation. Since there are often anastomotic branches between the pulmonary vessels and the vertebral veins, they are more likely to metastasize to the brain. The lung is an active organ. Changes in pleural pressure, coughing, and other factors can cause cancer cells to enter the blood circulation, which is related to BM.

The typical clinical course of lung cancer brain metastasis is that patients first develop symptoms of primary cancer, such as cough, hemoptysis, chest pain, etc., and neurological symptoms appear later. Some patients may have neurological symptoms or both at the same time, and a small number of patients only show neurological symptoms. The length of the interval between the diagnosis of primary cancer and BMs is related to the disease stage, pathological type, general condition of the patient, treatment, and treatment effect of the primary cancer. About 50% of the cases are diagnosed with primary tumor and BMs with the interval less than 1 year. More than 60% of BMs have clinical symptoms. The clinical manifestations of most patients with BMs are related to the tumor location and mass effect. Some patients have chronic symptoms such as headache. About 10% of patients have a sudden stroke-like episode as the first presenting symptom, mostly caused by intratumoral hemorrhage, tumor embolism, tumor necrosis, liquefaction, and cystic degeneration, which rapidly increases the tumor volume. Common symptoms include headache, vomiting, limb weakness, epilepsy, vision changes, aphasia, and ataxia. Headache is the most common symptom, and it is often the first symptom. About 50% of patients have headache as the early symptom. Headache can be caused by direct tumor invasion, compression, or increased intracranial pressure. Intracranial pressure increases due to tumor volume exceeding the compensatory capacity of the contents of the cranial cavity, cerebral edema, etc. The typical symptoms of increased intracranial pressure are the triad of headache, vomiting, and visual disturbance.

In addition, patients often have meningeal irritation, manifested as neck pain. The location of the headache is often the same as the location of the metastases, and localized headache has localization value, but it is not reliable. When the early intracranial pressure is still in the compensable stage, there may be only dizziness or mild headache, which lasts from a few minutes to several hours, and may be aggravated at night or in the morning, which can be relieved spontaneously. As the disease progresses, the headache becomes more obvious and persistent. Headache is aggravated by squatting, straining, talking, coughing, and defecation. When intracranial hypertension develops further, there will be inattention, stupor, lethargy, disorientation, and possibly focal symptoms, and even dystonia.

About 30% of patients with BMs are accompanied by neurological disorders, which are more common in patients with metastatic tumors involving fronto-temporal lobe or with extensive brain edema. The initial stage of BMs may only be manifested as dizziness, apathy, memory loss, unresponsiveness, emotional instability, photophobia, fear of noise, irritability, decreased alertness, and loss of orientation. Some patients have anxiety, personality changes, and so on. Later, drowsiness, lethargy, coma, and other disturbances of consciousness occur. Epilepsy is the first symptom in 10% of patients with BMs, which can be localized epilepsy, temporal lobe epilepsy, or generalized epilepsy. Seizures are especially common in supratentorial metastases. It should be noted that neurological symptoms in patients with lung cancer may also be caused by ectopic endocrine. The presentation of focal symptoms depends on involved intracranial structures and is characterized by a gradual, progressive loss of neurological function. Focal symptoms are common after the symptoms of increased intracranial pressure appear, but about 25% of patients have focal symptoms before the symptoms of increased intracranial pressure, especially for patients with frontal and parietal metastases. About 40% of patients with BMs have muscle weakness, such as slightly weaker upper limb grip strength on one side, lower limb weakness, etc. These symptoms are mild and easily overlooked by patients or their families and even clinicians. Hemiplegia mostly occurs in patients with supratentorial BMs, and it can also be caused by edema of brain tissue adjacent to tumor. Therefore, some cases can recover quickly after application of adrenal cortex hormones. Infratentorial metastases are more common with cerebellar symptoms such as nystagmus and ataxia. In addition, there may be cranial nerve involvement and other symptoms.

The patient's primary lung cancer history, clinical symptoms, and related imaging studies as CT or MRI to identify intracranial space-occupying lesions and lung cancer BMs can be established. BMs mostly occurred within 1 year after treatment. Patients with a history of lung cancer who have increased intracranial pressure and neuropsychiatric symptoms should be alerted to the diagnosis of intracranial metastases. In the diagnosis and treatment of lung cancer, attention should be paid to check for clinical symptoms of intracranial metastasis and performing corresponding neurological examinations. If there is any abnormality, further CT or MRI scans should be performed. In addition, for some patients who were misdiagnosed as primary brain tumors but were pathologically diagnosed as metastatic cancer after surgery, further examinations should be performed to find primary lesions, especially the lungs. The diagnosis of BMs mainly relies on imaging. CT and MRI are commonly used to determine the size, number, location and extent of BMs, and are used for evaluation of treatment response and complications.

CT is one of commonly used modalities for diagnosing BMs. Injecting contrast agent to enhance scanning can make the lesion clearer, and can clearly show the size, location, and number of metastases. MRI currently is the best choice for diagnosing BMs. Compared with CT, MRI has the advantages of better soft-tissue contrast in a multiplanar and multi-directional display, and can better distinguish intracranial anatomical structures. MRI is easier to detect BMs than CT, and can show early metastases even though no abnormality on CT. MRI can better show multiple lesions. The infratentorial metastases are easier to diagnose on MRI than on CT. Other examinations, such as stereotactic needle biopsy, are invasive, but can accurately puncture the tumor site under the imaging guidance. Radioisotope also has some value for the diagnosis of intracranial tumors. When diagnosing lung cancer BMs, attention should be paid to the identification of primary brain tumors, brain abscesses, and cerebrovascular diseases, so as not to be misdiagnosed as metastases. With neurological symptoms as the first manifestation, imaging revealed a single space-occupying lesion, and primary brain tumor should be excluded. Stereotactic needle biopsy or postoperative pathological or cerebral angiography is sometimes required to make a definite diagnosis.

CT findings (Fig. 4.3.1): ① Often located in the subcortical area of the cerebral hemisphere, 60%–70% cases show multiple lesions. ② Noncontrast CT scan is iso-density, low density, or high density. If it is a large low-density lesion, the outer wall of iso-dense is often seen. ③ The lesions showed moderate to obvious enhancement on enhanced scan, and the enhancement of multiple nodules was relatively uniform. Larger lesions could be enhanced in rings with thick and inhomogeneous walls. ④ Edema around the tumor can be extensive, especially in the parietal lobe. The tumor is small and the edema is large. ⑤ Mass effect is obvious. MRI findings (Fig. 4.3.1): ① The lesions are more common in the gray–white matter junction, and can also be confined to the white matter of the brain, often multiple. ② The lesions have various shapes, which can be nodular or cystic and solid. ③ The peritumoral edema is obvious; however, there is usually no edema around the small nodules below 4 mm. ④ After enhancement, it shows nodular, annular, or punctate enhancement. The borders and surroundings are clearly demarcated. When it presents as a single mass, it should be differentiated from glioma. When it presents as a single or multiple small nodule, it should be differentiated from granulomatous lesions; when the lesion presents as annular enhancement, it should be differentiated from brain abscess, glioblastoma, etc.

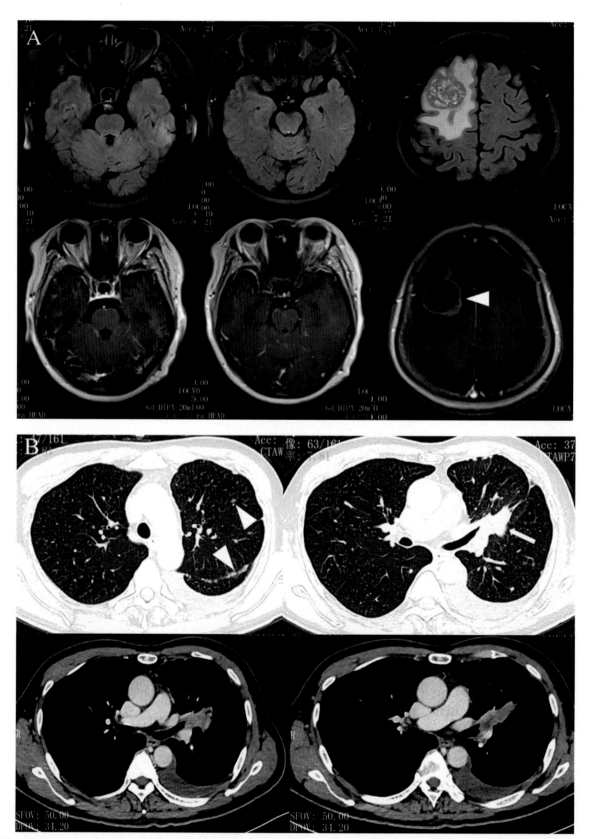

**FIGURE 4.3.1** A 61-year-old man with lung cancer with intrapulmonary, pleural, and brain metastases (A) Brain MRI showing brain metastases as both ring-enhancing lesions and leptomeningeal linear enhancement (*arrow*). (B) Chest CT showing lung cancer in the left upper lobe (*white arrow*) with multiple metastases and pleura metastases (*arrow*).

# References

[1] Tadros S, Ray-Chaudhury A. Pathological features of brain metastases. Neurosurg Clin October 2020;31(4):549—64.

[2] Achrol AS, Rennert RC, Anders C, et al. Brain metastases. Nat Rev Dis Prim January 17, 2019;5(1):5.

[3] Ostrom QT, Wright CH, Barnholtz-Sloan JS. Brain metastases: epidemiology. Handb Clin Neurol 2018;149:27—42.

[4] Page S, Milner-Watts C, Perna M, et al. Systemic treatment of brain metastases in non-small cell lung cancer. Eur J Cancer 2020;132:187—98.

[5] Olmez I, Donahue BR, Butler JS, et al. Clinical outcomes in extracranial tumor sites and unusual toxieities with concurrent whole brain radiation (WBRT) and Erlotinib treatment in patients with non-small cell lung cancer (NSCLC) with brain metastasis. Lung Cancer 2010;70(2):174—9.

[6] Preusser M, Capper D, Ilhan-Mutlu A, et al. Brain metastases: pathobiology and emerging targeted therapies. Acta Neuropathol 2012;123(2):205—22.

[7] Barnholtz-Sloan JS, Sloan AE, Davis FG, et al. Incidence proportionsof brain metastases in patients diagnosed (1973 to 2001) in the metropolitan detroit cancer surveillance system. J Clin Oncol 2004;22(14):2865—72.

Chapter 4.4

# Neurogenic pulmonary edema

Neurogenic pulmonary edema (NPE) refers to sudden pulmonary edema caused by craniocerebral injury or CNS pathology in the absence of primary cardiac, pulmonary, or renal disease, also known as central pulmonary edema. A variety of serious cranial neurological diseases including cerebrovascular disease, craniocerebral injury, epilepsy, and intracranial surgery can cause NPE, the most common of which is subarachnoid hemorrhage caused by ruptured aneurysm, followed by cerebral hemorrhage [1]. In addition, NPE is more difficult to treat and has a higher mortality rate, which predisposes to pulmonary infection, leading to impaired oxygen diffusion in the lungs and subsequent hypoxemia [2]. At present, the mechanism for the occurrence and development of NPE is not completely clear, mainly including the hemodynamic theory, the pulmonary capillary permeability theory and shock injury theory, etc. It is mostly believed that the sudden increase in intracranial pressure causes sympathetic excitation, which leads to changes in the dynamics and permeability of the pulmonary circulation [3]. Clinical features of NPE include a series of oxygenation dysfunction manifestations, such as dyspnea, shortness of breath, tachycardia, cyanosis, coughing pink foamy sputum, and popping sounds and rales on auscultation of both lungs. Hypoxemia is diagnosed mainly by a reduced partial pressure of oxygen, PaO2/FiO2 <200. There are two clinical subtypes of NPE, acute and chronic. The acute type presents with onset minutes to hours after neurological damage, commonly 30−60 min, while the chronic type presents with onset 12−24 h after neurological damage [4,5]. NPE usually resolves within 48−72 h after onset [6], and in some patients, the symptoms have subsided even before being examined. However, if the CNS lesion persists and the high cranial pressure does not resolve, the NPE will progressively worsen and persist.

NPE usually occurs within a few hours to a few days after severe cerebrovascular disease. NPE is usually diagnosed by chest X-ray and CT examination. Chest CT mainly shows patchy or large lamellar exudate, the lesions vary in size, and are not distributed according to lobar segments, but mainly by gravity distribution. It is similar to cardiogenic pulmonary edema and the lesions are mainly distributed on the dorsal side of both lungs, and the ventral side of both the lungs is basically unaffected. There is a certain gravity effect, called "gravity sign"; the lesions occurring in the posterior segment of the upper lobe are bounded by oblique fissures, which can clearly show the lower part of the lesions. When the lesions are near the oblique fissure and the dorsal pleura are slightly widened and thickened, it is called the "pleural pendant sign" [7] (Fig. 4.4.1).

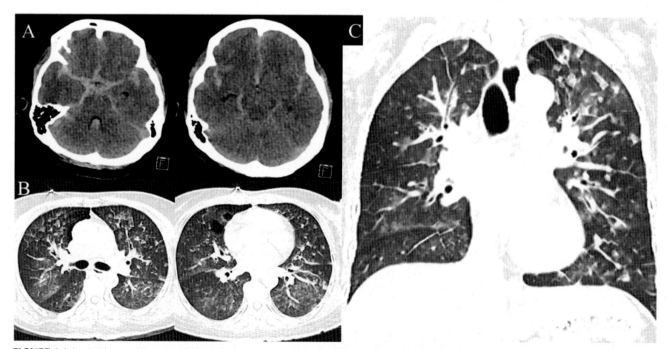

**FIGURE 4.4.1** A 44-year-old woman, 3 days after spontaneous subarachnoid hemorrhage. (A) CT showing subarachnoid hemorrhage. (B and C) Chest CT showing patchy exudation in both lungs, with lesions of different sizes and diffuse distribution.

# References

[1] Pyeron AM. Respiratory failure in the neurological patient: the diagnosis of neurogenic pulmonary edema. J Neurosci Nurs 2001;33(4):203−7.

[2] Fontes RB, Aguiar PH, Zanetti MV, et al. Acute neurogenic pulmonary edema: case reports and literature review. J Neurosurg Anesthesiol 2003;15:144−50.

[3] Meaudre E, Polycarpe A, Pernod G, et al. Contribution of the brain natriuretic peptide in neurogenic pulmonary oedema following subarachnoid haemorrhage. Ann Fr Anesth Reanim 2004;23(11):1076−9.

[4] Ana SC, Sónia M, Maria S. Neurogenic pulmonary edema due to ventriculo-atrial shunt dysfunction: a case report. Rev Bras Anestesiol 2016;66(2):200−3.

[5] Davison DL, Chawla LS, Selassie L, et al. Neurogenic pulmonary edema: successful treatment with IV phentolamine. Chest 2012;141(3):793−5.

[6] Tan CK, Lai CC. Neurogenic pulmonary edema. Can Med Assoc J 2007;177(3):249−50.

[7] Gong YJ, Guo JZ, Niu JJ, et al. Chest CT imaging characteristics of neurogenic pulmonary edema. Pract J Card Cerebr Pneumal Vasc Dis 2020;28(10):118−20.

Chapter 4.5

# *Nocardia* infection

*Nocardia* is a gram-positive aerobic bacterium, which can be infected by inhalation through the respiratory tract or by direct contact with injured skin [1]. *Nocardia* infection can present as an acute and chronic purulent or granulomatous disease, mainly seen in immunodeficient or immunosuppressed patients [2]. Pulmonary *Nocardia* is the most common seen nocardial infection (about 75%), and can be disseminated hematogenously to the CNS, kidney, and other parts of the body, forming multiple abscesses. *Nocardia* is an opportunistic pathogenic bacterium, not the normal flora of the body, and generally does not present endogenous infection. Factors that increase the risk of disease include chronic lung disease, cirrhosis, lymphatic system malignancies, solid organ, and bone marrow or stem cell transplantation, long-term gluco-corticoid use or Cushing syndrome, systemic lupus erythematosus, systemic vasculitis, ulcerative colitis, and other diseases. Occupational exposure may also be a risk factor for infection, such as agriculture, construction, and other environments where soil and dust are prevalent, increasing the probability of inhalation and traumatic exposure. In recent years, due to the widespread use of adrenocorticosteroids and immunosuppressants, the incidence of nocardial infection is on the rise.

Nocardial infection can spread to the whole body, and about 75% have the lung as the primary lesion. Clinical manifestations are not specific, mostly with fever, cough, sputum, chest tightness, etc. Involvement of the pleura may cause chest pain, and about 1/3 will develop pleural effusion, of which purulent pleural effusion is the most common, and 10% −40% of patients with *nocardia* will develop pus chest. Very rarely, a fistula is formed with the bronchus or chest wall, thus forming a pneumothorax. Extrapulmonary infection with *Nocardia* usually results from hematogenous dissemination of the primary lesion in the lung, and in about 1/3 of patients, it invades the brain, skin, eyes (keratitis), heart valves, liver, spleen, adrenal glands, and thyroid, forming migratory abscesses. The brain is the most common site of secondary infection, and intracranial *Nocardia* can cause meningitis and brain abscess.

Chest CT shows nonspecific findings, with patchy and solid opacifies most common seen, followed by nodular and cavernous cavity. The more specific findings can be [3−5]: (1) confined or diffuse pulmonary infiltrates. It usually shows solid lung lobules, and the whole lung lobe has high-density appearance after the fusion of adjacent lobular lesions; the involvement of interstitial lung showing a thin ground glass-like shadow. Most of the lesions are predominantly solid, often widely distributed, and bronchial inflation can be observed in some areas of solid lesions. (2) Single or multiple nodules and masses: nodules vary in size and can be corn-like nodules, most of which are large nodules and easily form cavities. The nodules may have low density, and there may be "halo signs" around the nodules, similar to fungal infection. (3) Cavitation: cavitation is more common and can occur in solid areas, nodules, or within masses. Cavitary lesions are more often seen in immunocompromised patients. (4) When the pleura is involved, pleural effusion may occur, which may also manifest as a pustular chest. Chronic pleural inflammation may manifest as pleural thickening and the chest wall muscles may become edematous and lose their normal fat density. (5) Occasionally, enlarged mediastinal and hilar lymph nodes may be seen.

Brain abscesses and meningitis are most common. About 1/3 of patients present with multiple intracerebral abscesses, and granulomas or diffuse involvement of the brain are also seen. The imaging shows more typical manifestations of intracranial infection. CT scan shows single or multiple patchy hypodense foci in the brain with a peripheral edematous band and mild mass effect, and the enhanced scan shows circular enhancement. On MRI, the lesion has equal or slightly low signal on the T1WI sequence and equal signal in the cyst wall, and high signal inside the T2WI and low signal in the cyst wall. The center of *Nocardia* brain abscess consists of a viscous fluid with germs, inflammatory cells, mucin, and cellular debris, especially the cellular density and viscosity of the pus center are high, which limits the diffusion of water molecules; therefore, the DWI shows high signal and low ADC value (Fig. 4.5.1) [6,7]. Lack of specificity of imaging presentation of *Nocardia* intracranial infection makes it more difficult to differentiate it from brain abscesses caused by other bacteria and fungi. However, this disease should be alerted to if refractory, septic, or mass lesions of the brain are encountered in clinical practice, especially in patients with underlying immunocompromised disease.

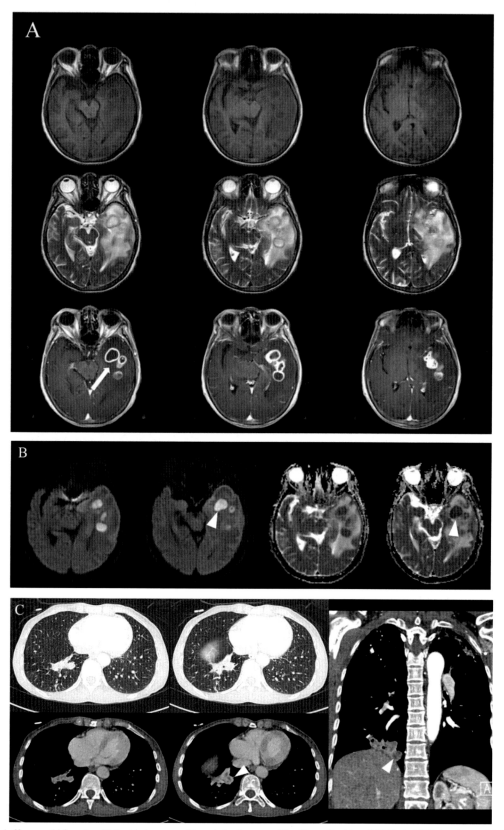

**FIGURE 4.5.1**   A 63-year-old female with brain abscess and lung infection caused by *Nocardia*. (A and B) Brain MRI showing multiple rim-enhancing lesions (*white arrow*) with restricted diffusion (*arrow*). (C) Chest CT show hypodense nodules coalesced into a mass in the right lower lobe on the postcontrast images (*arrow*).

# References

[1] Hemmersbach -Mille R, Catania J, Saullo JL, et al. Updates on nocardia skin and soft tissue infections in solid organ transplantation. Curr Infect Dis Rep 2019;21(8):27.

[2] Trivedi DP, Bhagat R, Nakanishi Y, et al. Granulomatous thyroiditis: a case report and literature review. Ann Clin Lab Sci 2017;47(5):620—4.

[3] Liu B, Zhang Y, Gong J, et al. CT findings of pulmonary nocardiosis: a report of 9 cases. J Thorac Dis 2017;9:4785—90.

[4] Kanne JP, Yandow DR, Mohammed TL, et al. CT findings of pulmonary nocardiosis. AJR Am J Roentgenol 2011;197:W266—72.

[5] Sato H, Okada F, Mori T, et al. High-resolution computed tomography findings in patients with pulmonary nocardiosis. Acad Radiol 2016;23:290—6.

[6] Pyatigorskaya N, Brugieres P, Hodel J, Mekontso Dessap A, Gaston A. What is your diagnosis? Nocardia abscessus infection. J Neuroradiol 2010;37:192—5.

[7] Asano M, Fujimoto N, Fuchimoto Y, Ono K, Ozaki S, Kimura F, et al. Brain abscess mimicking lung cancer metastases; a case report. Clin Imag 2013;37:147—50.

Chapter 4.6

# Tuberculosis

Tuberculosis is an infectious disease caused by *Mycobacterium tuberculosis* complex [1], which can occur in a variety of organs throughout the body, with the lungs being the most common involved. CNS tuberculosis accounts for 1.8%−5% of tuberculosis infections and 5%−15% of extrapulmonary tuberculosis. The disease can occur at any age. In China, it is mainly seen in children and adolescents, while in developed countries, it mainly involves adults, and patients are usually associated with immunodeficiency [2]. The pathological changes of tuberculosis mainly include exudative, proliferative, and necrotic lesions. During the development of tuberculosis, the above three pathological changes are often mixed, and at different stages, one pathological change is predominant and transforms into another, depending on the virulence of *M. tuberculosis*, the number of infected bacteria, and immunity of the body [3,4]. The main clinical manifestations of tuberculosis involving the lungs are cough, hemoptysis, chest pain, dyspnea and fever, usually manifested as a low fever in the afternoon. The localized symptoms and signs of the CNS infection may appear, depending on the site of TB lesions, as combined meningitis with symptoms of meningeal irritation such as headache, vomiting, cervical tonicity, or high cranial pressure; progression of the disease with impaired consciousness or coma [4].

Patients with primary *M. tuberculosis* infection often have no imaging abnormalities. If significant infection occurs, it often shows solid shadows in the air spaces, involving the entire lung lobes. Lymph node enlargement often occurs in children with primary tuberculosis infection. Sometimes they may invade the hilar and mediastinal lymph nodes, especially in the right paratracheal region. Lymph node enlargement is rare in adults with primary tuberculosis, except in immunocompromised patients. The lymph nodes often show a central hypointense shadow on contrast-enhanced CT, suggesting necrosis [5−7]. Secondary tuberculosis may present as infiltrative, fibrous cavitary, caseous pneumonia, interstitial tuberculosis, etc. Radiographically, secondary tuberculosis is more typical, occurring in the apical posterior and dorsal segments of the upper and lower lobes of the lung, mostly presenting as infiltrative lesions, often with the formation of cavities, with variable old and new lesions in the form of small patches or white patches, and with lesion borders poorly demarcated from surrounding normal tissue. The lesions show polymorphic changes such as patchy infiltrates, cavities, and nodular shadows of various sizes. Secondary tuberculosis with structural changes in the interstitial space of both upper lungs (mostly large patchy fusion lesions) is characterized by varying size of ground glass shadows (with low specificity) or lamellar distribution of fine reticulation lines with clear margins, but the structural contours of the lobular core are not easily recognizable, and may be accompanied by diffuse distribution of microscopic nodular shadows with poorly defined margins. It may be accompanied by diffusely distributed tiny nodular shadows with clear or blurred margins, 1−2 mm in diameter, and may form a tree-bud sign if connected with tiny branching shadows occurring in the lobular core. In severe cases, large lamellar solid shadows may be formed, and cavitation may be seen within. In the lower lung, the lesions are typically diffuse, small flaky tuberculosis lesions, which may be white snowflakes with distinctive shapes. When a cavity is present in a lesion, a thin-walled cavity with a wall thickness of <3.0 mm is considered a thin-walled cavity, and a thick-walled cavity with a wall thickness of ≥3.0 mm is common in active tuberculosis; a cavity with a diameter of less than 2.0 cm and a lobular or nodular shape outside the cavity is called a small focal cavity if a cavity with a uniform wall appears inside the cavity. The formation of cavities is mostly due to necrosis or liquefaction of the caseous-like material in the lesion, which is discharged from the body or absorbed by the body, and the lesion becomes empty to form a cavity, and about 40% of the cavities can appear in the air−fluid plane. The wall-less cavity is a sign of caseous necrosis of the material in the lesion, which is caused by coughing up the necrotic material in the cavity through the draining bronchus during coughing, with smooth walls, and the draining bronchus can be seen in most cavitated lesions. Enhanced CT examination reveals isolated nodules or solid shadows, and the degree of enhancement of tuberculous lesions needs to be focused on with time-lapse scanning of enhanced CT. By time-lapse scanning of the lesion, the lesion area shows heterogeneous or mild enhancement, which can be accompanied by multiple hypointense lesions with more marginal enhancement [8].

CNS tuberculosis is divided into tuberculous meningitis, parenchymal tuberculosis, and mixed intracranial tuberculosis according to the site of involvement [2]. Regarding the tuberculous meningitis, the density of the subarachnoid space is increased on CT noncontrast scan, obvious in the suprasellar space. Punctate calcification is seen in the later stages, and significant enhancement with irregular lesions is seen on enhanced scan. MRI shows a slightly long T1 and long T2 signal in the brain, and significant enhancement. Cerebral edema, hydrocephalus and signs of cerebral infarction may also be seen. Regarding the intracerebral parenchymal tuberculosis, CT scan shows equal or slightly hypointense, with enhancement. T2WI and FLAIR images show high signal, DWI shows equal or slightly high signal, and enhancement scan shows circumferential enhancement or nodular enhancement, and the ring wall is often bright and neat, and edema signal is seen around the foci. Regarding the intracerebral tuberculosis lesions, CT scan shows equal, high, or mixed density, with visible calcification and mild edema. When the center of tuberculoma has coagulative necrosis, T1WI shows high signal and T2WI shows low signal; when it is liquefied necrosis, MRI features are similar to that of abscess (Fig. 4.6.1).

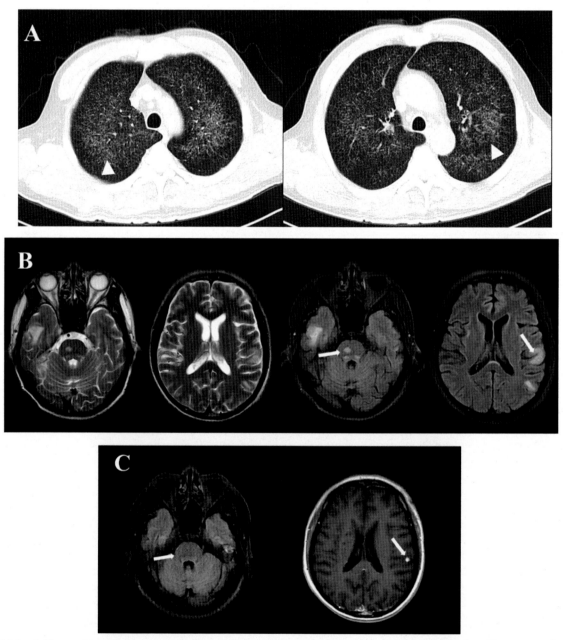

**FIGURE 4.6.1**   A 62-year-old man with acute hematogenous pulmonary tuberculosis. Figure (A) CT chest showing the miliary pattern of pulmonary micronodules in uniform size, density, and distribution throughout both the lungs (*arrow*). (B) Brain MRI showing scattered lesions with high signal intensity on T2WI and FLAIR in the brainstem and both cerebral hemispheres with associated edema (*white arrow*). (C) Follow-up brain MRI showing the lesions in the brainstem and the right temporal lobe resolved after antituberculosis treatment for 3 months, while lesions in the left temporal lobe persist (*white arrow*).

# References

[1] Ding C, Hu M, Shangguan Y, et al. Epidemic trends in high tuberculosis burden countries during the last three decades and feasibility of achieving the global targets at the country level. Front Med March 3, 2022;9:798465.

[2] Rodriguez-Takeuchi SY, Renjifo ME, et al. Extrapulmonary tuberculosis: pathophysiology and imaging findings. Radiographics 2019;39:2023−37.

[3] Pere-Joan C. Pathogenesis of tuberculosis and other mycobacteriosis. Enferm Infecc Microbiol Clín 2018;36(1):38−46.

[4] García JR, González A. Tuberculosis: pathogenesis and clinical manifestations. Acta Cient Venez 2001;52(Suppl. 1):5−9.

[5] Wormanns D. Radiological imaging of pulmonary tuberculosis. Radiologe 2012;52(2):173−84.

[6] Nachiappan AC, Rahbar K, Shi X, et al. Pulmonary tuberculosis: role of radiology in diagnosis and management. Radiographics 2017;37(1):52−72.

[7] Alvarez Martín T, Merino Arribas JM, Ansó Oliván S, et al. Clinical and radiological characteristics of primary pulmonary tuberculosis in adolescents. An Esp Pediatr 2000;52(1):15−9.

[8] Tanaka D, Niwatsukino H, Oyama T, Nakajo M, et al. Progressing features of atypical mycobacterial infection in the lung on conventional and high resolution CT (HRCT) images. Radiat Med 2001;19(5):237−45.

Chapter 4.7

# Tuberous sclerosis complex

Tuberous sclerosis complex (TSC), also known as Bourneville's disease, is an autosomal dominant neurocutaneous syndrome with skin lesions, seizures, and mental retardation as the main clinical features, with an incidence of 1 per 100,000, which can be familial or disseminated, and more males than females. It can involve multiple organs such as the skin, brain, kidney, liver, heart, lung, retina, and bone [1]. The disease is mainly caused by mutations in the tumor suppressor genes TSC1 and TSC2, of which TSC2 mutations are more common and about 15% of patients do not have mutations [2]. The proteins encoded by TSC1 and TSC2 form a complex that indirectly inhibits cell growth and protein synthesis, and when mutated, this inhibitory function is disrupted, leading to abnormal cell proliferation and mismatch formation. In general, this abnormal proliferation is limited and does not lead to malignant transformation. Most clinical features of TSC become apparent only after the age of 3 years [3], with epilepsy, mental retardation, and sebaceous adenoma of the face (Vogt's triad), as its characteristic clinical manifestations. CNS abnormalities result in the highest morbidity and mortality in patients with TSC, and the main pathological change is a neuroglial proliferative sclerosing nodule that occurs widely in the cerebral cortex, white matter, and subventricular space, which can form subventricular nodules, cortical nodules, cerebral white matter abnormalities, and subventricular giant cell astrocytomas [2]. Pulmonary features of TSC include lymphangioleiomyomatosis (LAM) in cysts and nodules in multifocal micronodular pneumocyte hyperplasia (MMPH) [2]. LAM is the main pulmonary manifestation of TSC in 30%–40% of female patients with TSC, usually presenting with progressive dyspnea and recurrent pneumothorax at the age of 30–40 years. MMPH is the result of type II alveolar cell proliferation.

The main types of intracerebral involvement of TSC are listed as follows: (1) Subventricular multiple nodules (Figs. 4.7.1 and 4.7.2): It is an important manifestation of TSC, mainly composed of giant cell masses, the nodules are distributed along the lateral ventricular wall and protrude into the ventricles, CT shows isointense or calcification, or slightly hypointense, shaped like teardrops; T1WI and T2WI show iso- or low-signal nodules on MRI, with no enhancement on postcontrast scan. (2) Intracortical nodules: they may involve the cerebral or cerebellar cortex, with iso- or slightly hypointense or calcified foci on CT, without surrounding edema or mass effect. MRI may show swelling of the cerebral gyrus, and larger nodules show giant cerebral gyrus, with unclear boundaries in adjacent cortex and white matter, with slightly low or low signal on T1WI and high signal on T2WI and FLAIR, with no enhancement on postcontrast scan [4]. (3) Abnormal signal in white matter (Fig. 4.7.3): linear shadows extending from the ventricle or paraventricular white matter to normal cortex or cortical nodules are the most common findings. These lesions can also be expressed as wedge-shaped lesions or spherical lesions with tips pointing to ventricular wall. (4) Subependymal giant cell astrocytoma evolves

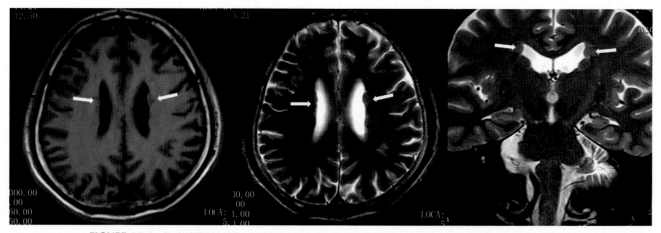

**FIGURE 4.7.1**   Brain MRI showing multiple subependymal nodules along the lateral ventricles (*white arrows*).

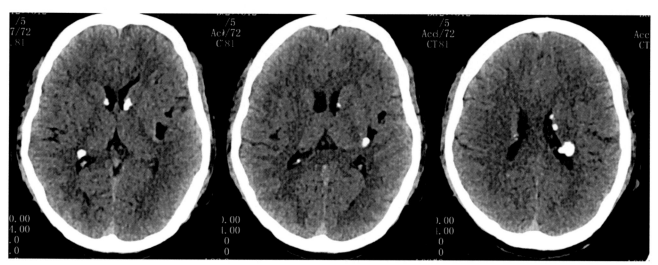

**FIGURE 4.7.2**   Non-contrast CT brain shows multiple calcified subependymal nodules along the lateral ventricles.

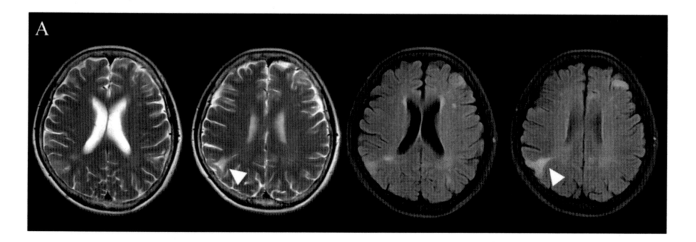

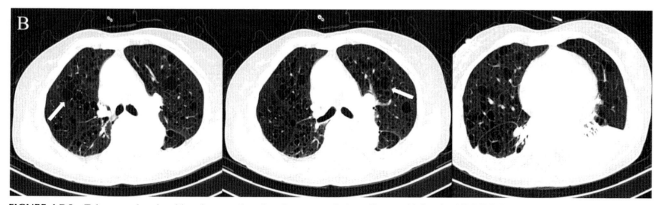

**FIGURE 4.7.3**   Tuberous sclerosis with pulmonary lymphangiomyomatosis in a 48-year-old female. (A) MRI showing wedge-shaped hyperintense lesions in cerebral cortex with the tip pointing toward ventricle wall (*arrow*). (B) Chest CT showing diffuse thin-walled cysts (*white arrow*).

from subventricular sclerotic nodules [5]. The tumor is often located near the foramen magnum, triangle, or third ventricle of lateral ventricle, and the larger lesions may cause hydrocephalus (Fig. 4.7.4).

Chest CT showed hypodensity and enlargement of both lungs with emphysema-like changes (Fig. 4.7.3). High-resolution computed tomography (HRCT) clearly shows the characteristic widely distributed cystic shadow in both the

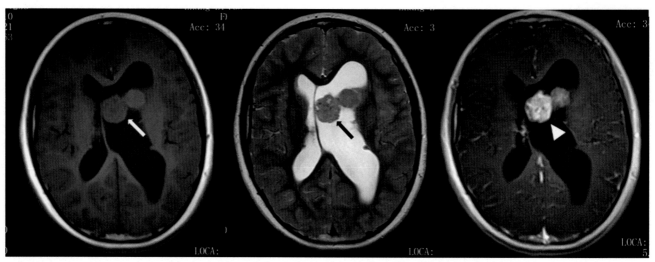

**FIGURE 4.7.4** Subependymal giant cell astrocytoma in an 8-year-old girl. The lesions located in the body of the left ventricle are iso-intensity on T1WI (*white arrow*) and hyperintensity on T2WI (*black arrow*), and inhomogeneous enhancement on postcontrast MRI (*arrow*).

lungs with diffuse distribution pattern. The diameter of cystic shadow is between 2 and 20 mm, mostly less than 10 mm, and the wall of cyst is mostly less than 3 mm, with relatively normal tissue between walls and blood vessels located around cystic cavity. As disease progresses, cystic shadow tends to increase in size, with partial fusion leading to interstitial fibrosis [6]. MMPH appears on CT as diffusely distributed small nodular hyperdense shadows with lobular septal thickening, with nodule size of 1−10 mm [7].

# References

[1] Northrup H, Krueger DA. Tuberous sclerosis complex diagnostic criteria update: recommendations of the 2012 international tuberous sclerosis complex consensus conference. Pediatr Neurol 2013;49(4):243−54.

[2] Curatolo P, Bombardieri R, Jozwiak S. Tuberous sclerosis. Lancet 2008;372(9639):657−68.

[3] Korf BR. Neurofibromatosis. Handb Clin Neurol 2013;111:333−40.

[4] Christophe C, Sekhara T, Rypens F, et al. MRI spectrum of cortical malformations in tuberous sclerosis complex. Brain Dev 2000;22(8):487−93.

[5] Grajkowska W, Kotulska K, Jurkiewicz E, et al. Brain lesions in tuberous sclerosis complex. Review. Folia Neuropathol 2010;48(3):139−49.

[6] Avila NA, Dwyer AJ, Rabel A, et al. Sporadic lymphangioleiomyomatosis and tuberous sclerosis complex with lymphangioleiomyomatosis: comparison of CT features. Radiology 2007;242(1):277−85.

[7] Wang MX, Segaran N, Bhalla S, et al. Tuberous sclerosis: current update. Radiographics 2021;41(7):1992−2010.

# Chapter 5

# Cardiovascular system

Jing Wang and Teng Jin

*Department of Radiology, Union Hospital, Huazhong University of Science and Technology, Wuhan, China*

## Chapter 5.1

# Neurological complications of cardiac surgery

Jing Wang

Cardiac surgery may cause neurological complications that seriously affect the prognosis. With the development of coronary artery bypass grafting (CABG) and heart transplantation, the incidence of neurological complications has increased. Patients may suffer from stroke, hypoxic ischemic encephalopathy, etc. Some patients may also experience cognitive impairment. In addition, the brachial plexus and phrenic nerves may also be injured during cardiac surgery. There are multiple risk factors for encephalopathy after cardiac surgery including preoperative stroke, hypertension, diabetes, hyperlipidemia, advanced age, carotid artery stenosis, excessive alcohol consumption, intraoperative hypoxia and hypoperfusion, and postoperative atrial fibrillation [1]. The underlying mechanism for neurological complications after cardiac surgery includes embolism and hypoperfusion. Prior studies suggested brain injury after cardiac surgery is caused by embolism of gas and compound particles [2]. Other studies have shown that brain injury may be due to abnormal cerebral perfusion before surgery [2].

The main complications of brain injury after cardiac surgery are stroke, seizures, and coma. Studies have shown that stroke occurs in 1%−5% of patients after CABG. Most stroke patients do not recover fully and remain disabled. Stroke occurs 2−3 days after CABG, and the incidence of stroke may increase within 2 weeks after surgery [3]. Stroke rates are higher when heart valve surgery is combined with CABG [1]. Acute brain injury after cardiac surgery exists on a broad spectrum ranging from major stroke to subclinical brain injury, which includes postoperative cognitive dysfunction (POCD) and silent brain infarcts [4]. According the Trial of ORG 10172 in Acute Stroke Treatment criteria [5], ischemic stroke can be divided into large artery atherosclerosis, cardio-embolism, small artery occlusion, other unusual determined etiologies (OC), and stroke of undetermined etiology (SUD). OC and SUD are the main types of stroke after cardiac surgery. Brain MRI, especially diffusion-weighted imaging (DWI) is the preferred technique that can define acute stoke, which can demonstrate small ischemic lesions as bright hyperintensities that are evident within a few hours of onset of ischemia and generally disappear within 14 days. DWI combined with apparent diffusion coefficient can differentiate acute and chronic infarcts (Fig. 5.1.1). Neuroimaging patterns of stoke include major territory embolism, multifocal embolism characteristic of a proximal source of embolism, and a multifocal border zone pattern best detected with DWI. DWI and perfusion-weighted imaging can demonstrate the stage and degree of ischemic, magnetic resonance angiography (MRA) and digital subtraction angiography (DSA) can show the location and extent of vascular stenosis. It has been reported that intraarterial thrombolytic therapy is effective for stroke occurring 1−14 days after cardiac surgery [6].

Intracranial hemorrhage is not a common cause of stroke, but early detection is important to avoid delay of treatment. Intracranial hemorrhage may be related to the decrease of platelet adhesion function and the decrease of coagulation factors during cardiopulmonary bypass surgery [6]. Hematomas may locate in the brain parenchyma, subdural, or epidural space. Subdural hematomas can be presenting as postoperative seizures or focal neurologic deficit. The imaging features for

Multi-system Imaging Spectrum associated with Neurologic Diseases. https://doi.org/10.1016/B978-0-323-91795-7.00003-8

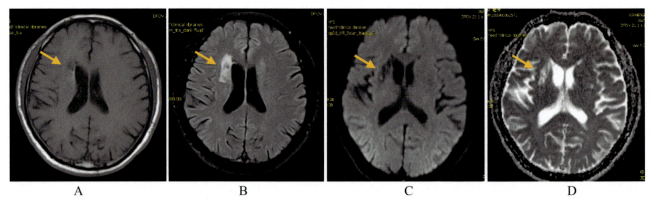

**FIGURE 5.1.1** Axial T1-weighted imaging (T1WI) (A), fluid-attenuated-inversion-recovery (FLAIR) imaging (B), Diffusion-weighted imaging (DWI) (C) and apparent diffusion coefficient (ADC) (D) of a patient with chronic infarction in right basal ganglia after cardiac valvular surgery.

**FIGURE 5.1.2** CT of a patient with multiple hemorrhages in left frontal and occipital lobes from infective endocarditis after mitral valve replacement, presenting large flaky homogeneous hyperdensity (arrows).

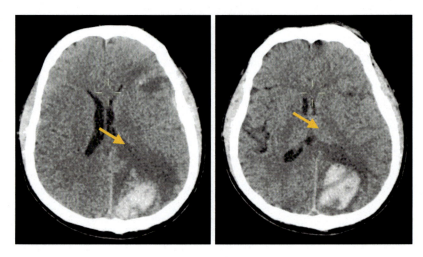

intracranial hemorrhage include both acute bleeding as well as acute on chronic bleeding on MRI [7]. CT is also a sensitive method for detecting intracranial hemorrhage presenting homogeneous hyperdensity (Figs. 5.1.2 and 5.1.3).

The manifestations of encephalopathy after cardiac surgery are diverse, such as confusion or impaired cognitive function, delirium, stupor, or coma. Patients with a neurological complication are slow to awaken from anesthesia, but stupor or coma is less common, occurring in less than 1% [8]. Coma may be caused by global ischemia and hypoxia, generalized stroke, or multiple lesions in the brain. For these patients, brain MRI is essential as it may be due to major territory cerebral infarction with a risk of life-threatening brain swelling that may warrant surgical intervention (Fig. 5.1.4). The sensitivity of DWI in identifying cerebral infarction is at least twice as better than that of CT. In addition, less common causes of encephalopathy or coma include hypoglycemia, a hypernatremic state, and acute obstructive hydrocephalus [6]. Seizures occur in about 3% of patients following cardiac surgery and are often indicative of an underlying ischemic injury. Epilepsy after cardiac surgery can be accompanied by coma, encephalopathy, or delirium. The incidence is less than 1%, and usually occurs in the early postoperative period within 24 h after surgery [9].

# 1. Peripheral nerve injuries

The brachial plexus and phrenic nerves are most common injured peripheral nerves during cardiac surgery. Polyneuropathy also occurs in some cases. Sternal retraction and internal mammary artery dissection are key factors responsible for brachial plexus neuropathy [10]. Direct trauma of the brachial plexus can occur by first rib fracture fragments or the associated fracture hematoma directly compressing the nerves. Persistent brachial plexus injury has been reported in 5% of patients after median sternotomy. Transient and minor brachial plexus injuries may be more common. Injury to the nerve from the lower trunk of the arm is most common, involving the triceps, resulting in

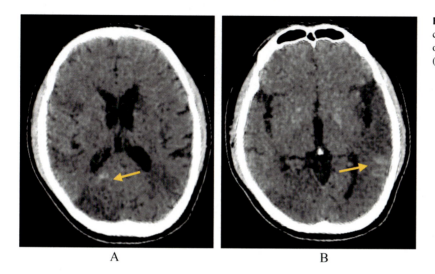

**FIGURE 5.1.3** Axial CT of a patient with multiple cerebral infarctions combined with hemorrhage in right occipital lobe (A, *arrow*) and left temporal lobe (B, *arrow*).

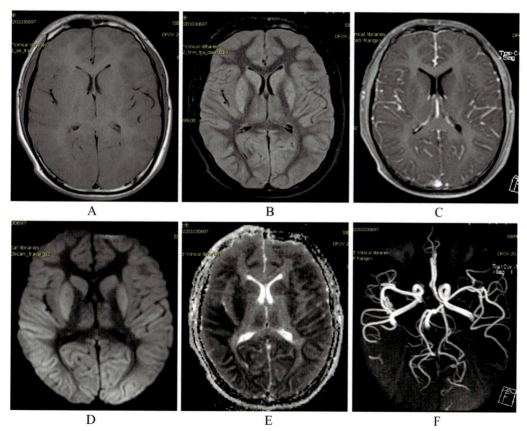

**FIGURE 5.1.4** Axial T1WI, FLAIR, contrast-enhanced T1WI (CE-T1WI), DWI, ADC, and MR angiography (MRA) of a patient with hypoxic-ischemic encephalopathy after cardiopulmonary resuscitation. (A) and (B). Axial T1WI and FLAIR showing extensive swelling in the cerebral cortex indicating diffuse brain edema. (C). Axial CE-T1WI showing no abnormal enhancement lesions. (D) and (E): DWI and ADC have no diffusion limitation. (F): TOF-MRA showing no obvious vascular stenosis and occlusion.

decreased reflexes, present with weakness and sensory loss in the ulnar C8-T1 distribution. Loss of sensation is sometimes present in the injured hand. Some patients may have pain, and a small number of patients may have Horner syndrome with miosis, ptosis, and anhidrosis. Intraoperative electrophysiological techniques can identify abnormal brachial plexus function, predict postoperative nerve injury, and determine what may induce brachial plexus injury during operation. With the emergence and continuous development of MR neurography, proper diagnosis has been provided for early identification of brachial plexus injury after cardiac surgery.

Open heart surgery can cause unilateral phrenic nerve damage resulting in unilateral diaphragmatic paralysis in at least 10% of patients, most commonly the left side [10]. Topical hypothermia and internal mammary artery dissection are the main factors that caused phrenic nerve damage. Manual manipulation and ischemia can cause phrenic nerve injury. Unilateral phrenic nerve injury can lead to atelectasis and respiratory muscle weakness and induce postoperative respiratory complications. Chest radiography, fluoroscopy, spirometry, ultrasonography, and transcutaneous phrenic nerve stimulation are the commonly used modalities to diagnose phrenic neuropathy. Studies have shown that 85% of patients with phrenic nerve injury have abnormal chest radiographs after surgery. Atelectasis was the most common chest radiographic finding, followed by diaphragm elevation. Bilateral diaphragmatic paralysis in cardiac surgery patients is very rare and requires high suspicion to diagnose, and can lead to death due to respiratory failure. The first indication of bilateral phrenic neuropathy may be difficulty in getting rid of mechanical ventilation. In this setting, chest radiographs are difficult to diagnose bilateral phrenic nerve injury, and measurement of transabdominal pressure using gastric and esophageal catheters may be the only diagnostic tool available for a definitive diagnosis [6].

Mononeuropathy due to intraoperative compression or trauma can involve the accessory, facial, lateral femoral cutaneous, peroneal, radial, recurrent laryngeal, saphenous, long thoracic, and ulnar nerves, resulting in unique clinical manifestations. Most compressive mononeuropathies are transient and usually reversible in 4–8 weeks.

## 2. Abnormal visual function

Visual disturbances after cardiac surgery are common, but are usually asymptomatic and reversible. Retinal infarction with cotton wool exudates occurs in 10%–25% of patients [6]. Retinal emboli are rare. But retinopathy is not associated with vision loss. Anterior ischemic optic neuropathy is an uncommon and disabling complication of cardiac surgery that results in infarction of the optic nerve head, papilledema, and painless permanent vision loss. Retrobulbar ischemic optic neuropathy due to intraorbital nerve injury occurs less frequently. It can cause acute blindness without papilledema, which can be accompanied by abnormal pupillary reflexes. Both anterior and posterior ischemic optic neuropathy can lead to unilateral or bilateral blindness. The reduced posterior ciliary artery blood flow in the setting of hypotension is the main mechanism, and some studies have also suggested postoperative anemia as a risk factor.

Cortical blindness due to bilateral occipital cortical ischemia manifests as ipsilateral visual field defect. Retinal and pupillary examinations can be normal in patients with cortical blindness. Some patients do not have any symptoms of visual impairment. There may be partial recovery from cortical blindness. Pituitary apoplexy is a rare complication of pituitary adenomas but its occurrence can be precipitated by recent cardiopulmonary bypass surgery. Patients may present with headache, ptosis, ophthalmoplegia, and visual impairment due to compression of adjacent cranial nerves and the anterior visual pathway.

## 3. Cognitive dysfunctions

POCD is a general complication following cardiac and major noncardiac surgery among the elderly, yet its causes and mechanisms are still unknown. Early POCD may be caused by embolism, hypoperfusion injury, blood–brain barrier damage caused by systemic inflammation, toxic metabolic encephalopathy, and delirium from being at intensive care unit. Preoperative brain arterial spin-labeling MRI may potentially predict the development of early cognitive dysfunction [11]. Alterations of preoperative cerebral blood flow might be a biomarker for early POCD. The presence of late cognitive decline means that some permanent brain damage has occurred in the perioperative period. One hypothesis is that diffuse microembolic cause neuronal damage, thereby accelerating brain atrophy and dementia. Another hypothesis is that the gradual development of vascular risk factors leading to chronic microvascular occlusion leading to cerebral ischemia is another cause of delayed cognitive decline after cardiac surgery (Fig. 5.1.5).

## 4. Heart transplantation

Heart transplantation is indicated in patients with progressive, end-stage heart failure with very poor cardiac function. The 1-year survival rate after transplantation is 80%–85%, the 5-year survival rate is 60%–80%, the 10-year survival rate is about 50%, and 15-year survival rate is 30%–40%. Intraoperative or postoperative neurological complications can be life threatening. The incidence of neurological complications is as high as 30%–80% [12]. Early detection and treatment of

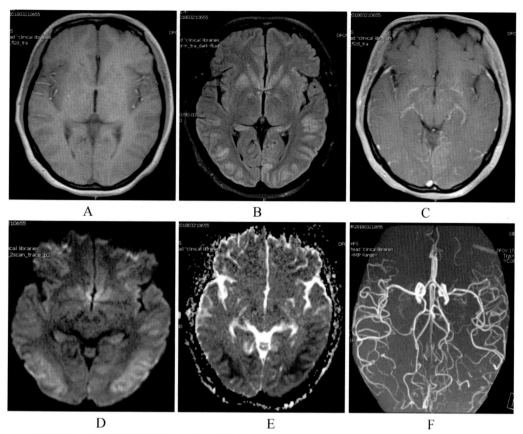

**FIGURE 5.1.5** Axial T1WI, FLAIR, CE-T1WI, DWI, ADC and MRA of a patient with cognitive dysfunction after cardiac arrest caused by myocarditis. (A−C). T1WI, FLAIR, and CE-T1WI showing slight swelling in temporal lobe and occipital lobe; (D) and (E). DWI and ADC showing diffusion limitation of subcortical white matter in temporal lobe and occipital lobe; (F): TOF-MRA showing no obvious vascular stenosis and occlusion.

complications are prudent. The neurological complications of heart transplantation are similar to those of heart valve or coronary artery bypass surgery, but the incidence is high. Central nervous system (CNS) complications after heart transplantation mainly include immunosuppressive-related neurotoxicity, epilepsy, stroke, encephalopathy, headache, mental disorders, peripheral neuropathy, CNS infection, and new CNS malignancy [12].

**(1)** To effectively reduce acute rejection after heart transplantation, individualized use of immunosuppressive agents is required. Commonly used immunosuppressants after heart transplantation are calcineurin inhibitors (CNIs), mycophenolate mofetil, glucocorticoids, etc. Among them, CNI is the main drug that causes postoperative immunosuppressant-related neurotoxicity, manifested as reversible posterior encephalopathy syndrome (PRES). The pathogenesis of PRES is the dysfunction of cerebrovascular regulation and the destruction of the integrity of the blood−brain barrier [13]. The lesions are more common in the back half of the brain, and a few involve the frontal and temporal lobes. Typical imaging findings are bilateral symmetric vasogenic edema in subcortical white matter areas of the parieto-occipital lobes, sometimes with restricted diffusion centers in PRES lesions, suggesting that vasogenic edema transforms into cytotoxic edema, resulting in irreversible damage (Fig. 5.1.6).

**(2)** Epilepsy occurs in about 15% of heart transplant patients, often during surgery. Epilepsy during transplant surgery usually occurs due to stroke. The most common type is generalized tonic-clonic seizure due to CNI-associated PRES. PRES signal features dominated by parieto-occipital involvement can be seen on MRI, presenting as reversible cerebral vasoconstriction syndrome (RCVS), vascular lumen stenosis, and "beaded" morphological features on MRA. Cerebral infarction or cerebral hemorrhage may also be seen [14].

**(3)** Clinical manifestations of encephalopathy after heart transplantation can vary from mild changes of consciousness to severe case such as coma. Metabolic disturbances and drug toxicity are the main causes for encephalopathy after heart transplantation. The postoperative encephalopathy includes PRES, RCVS, progressive multifocal leukoencephalopathy (PML), and osmotic demyelination syndrome (ODS). The differentiation of these encephalopathies is essential for

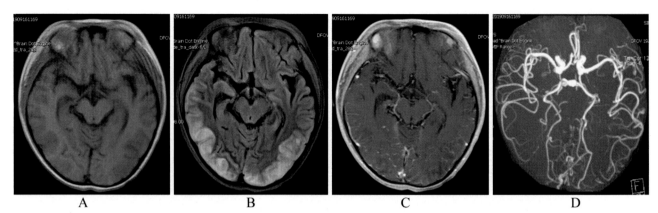

**FIGURE 5.1.6** Axial T1WI, FLAIR, CE-T1WI, and TOF-MRA of a patient with posterior encephalopathy syndrome (PRES). (A) and (B): extensive swelling in temporal lobe and occipital lobe with hyperintensity on FLAIR; (C): Multiple small vessels enhancement; (D): TOF-MRA showing no obvious vascular stenosis and occlusion.

treatment. The typical locations of PML are the subcortical white matter and the cerebellar peduncle. The lesions are irregular in shape and multiple bilaterally. The lesions of ODS are located in the center of the pons or outside the pons and often symmetrically distributed.

(4) Hallucinations, delusions, and schizophrenia may occur within 2 weeks or later after heart transplantation. If it occurs in recent postoperative period, which may be related to a variety of incidental etiologies, abnormal mental behavior usually decreases gradually over time. If it occurs later in the postoperative period, it may be related to intracranial infection, especially viral infection.

(5) Immunosuppression is also a possible late complication after heart transplantation. Opportunistic bacterial infection and immunosuppression can occur as early as 2 weeks after surgery. But it usually occurs at least one month after surgery. Focal meningoencephalitis or brain abscess, meningitis, and diffuse encephalitis are the three most common intracranial infections after heart transplantation. According to the source of infection, it can be divided into bacterial infection, fungal infection, viral infection and parasitic infection. *Aspergillus, Toxoplasma gondii, Cryptococcus neoformans, Listeria monocytogenes,* and herpes virus are common CNS infectious agents in patients with heart transplant. Aspergillosis is the most common fungal infection, manifesting as necrotizing meningoencephalitis, single or multiple brain abscesses. On MRI or CT, the abscess is annular and irregular in shape without enhancement. DWI may suggest limited diffusion of water molecules in the center of abscess. It can also invade blood vessels and cause ischemic and hemorrhagic cerebral infarction. Therefore, local hemorrhagic changes on brain imaging should consider *Aspergillus* infection. *Toxoplasma* infection is the second most common cause of focal or multiple meningoencephalitis and brain abscesses after heart transplantation. Contrast-enhanced CT showed multiple annular enhancements. MRI can find lesions that are not obvious on CT, and can also be used to quickly judge the efficacy of antibiotic therapy. Meningitis after heart transplantation is most commonly caused by *C. neoformans* infection, and its onset is subacute or chronic, and the white blood cell count in cerebrospinal fluid is slightly to moderately increased, mainly monocytes. The most common infection was *L. monocytogenes*. It presents as ring-like enhancement on postcontrast CT or MRI. Cytomegalovirus, herpes simplex virus, and shingles encephalitis infections can also occur in heart transplant patients, which are associated with dissemination of viremia. In the immunosuppressed state, acute focal necrotizing herpes simplex encephalitis can evolve into a chronic course. Imaging findings of the CNS due to viral infection lack specificity, which shows nonenhancement, diffuse, or gyrus-like enhancement on contrast-enhanced CT and MRI.

(6) Heart transplant recipients are at higher risk of developing newly malignant brain tumors. More than 10% of recipients develop malignant brain tumors within 1−5 years after transplantation. Posttransplant lymphoproliferative disorders (PTLDs) are common [15]. Early postoperative PTLD is associated with Epstein−Barr virus infection, whereas late PTLD is associated with long-term immunosuppression. The radiological patterns are varied on MRI, which needs to be differentiated from glioblastoma, primary CNS lymphoma and metastasis, etc.

# References

[1] Li JY, Bi Q. Neurological complications after coronary artery bypass grafting. Chin J Geriatr Heart Brain Vessel Dis 2012;14(5):552−3.

[2] Mills SA. Cerebral injury and cardiac operations. Ann Thorac Surg November 1993;56(5 Suppl. 1):S86−91.

[3] Merkler AE, Chen ML, Parikh NS, et al. Association between heart transplantation and subsequent risk of stroke among patients with heart failure. Stroke 2019;50(3):583−7.

[4] Indja B, Woldendorp K, Vallely MP, et al. Silent brain infarcts following cardiac procedures: a systematic review and meta-analysis. J Am Heart Assoc 2019;8(9):e010920.

[5] Wu X, Zou Y, You S, et al. Distribution of risk factors of ischemic stroke in Chinese young adults and its correlation with prognosis. BMC Neurol 2022;22(1):26.

[6] Sila C. Neurologic complications of cardiac surgery. Emerg Manage Neurocrit Care 2012.

[7] Rindler RS, Allen JW, Barrow JW, et al. Neuroimaging of intracerebral hemorrhage. Neurosurgery 2020;86(5):E414−23.

[8] Hrdlicka CM, Wang J, Selim M. Neurological complications of cardiac procedures. Semin Neurol 2021;41(4):398−410.

[9] Lin P, Tian X, Wang X. Seizures after transplantation. Seizure 2018;61:177−85.

[10] Sharma AD, Parmley CL, Sreeram G, et al. Peripheral nerve injuries during cardiac surgery: risk factors, diagnosis, prognosis, and prevention. Anesth Analg 2000;91(6):1358−69.

[11] Du X, Gao Y, Liu S, et al. Early warning value of ASL-MRI to estimate premorbid variations in patients with early postoperative cognitive dysfunctions. Front Aging Neurosci 2021;13:670332.

[12] Gu M, Liao ZK, Shi L, et al. Research progress of neurologic complications after heart transplantation. Chin J Transplant (Electronic Edition) 2020;14(4):255−7.

[13] Gewirtz AN, Gao V, Parauda SC, et al. Posterior reversible encephalopathy syndrome. Curr Pain Headache Rep 2021;25(3):19.

[14] Qubty W, Irwin SL, Fox CK. Review on the diagnosis and treatment of reversible cerebral vasoconstriction syndrome in children and adolescents. Semin Neurol 2020;40(3):294−302.

[15] Baldassari L E, Wattjes M P, Cortese I C M, et al. The neuroradiology of progressive multifocal leukoencephalopathy: a clinical trial perspective[J]. Brain J Neurol

Chapter 5.2

# Neurologic complication of infective endocarditis

Jing Wang

The relationship between infective endocarditis (IE) and arterial embolism was first discovered in the mid-19th century. In 1885, Osler first described the typical triad: fever, cardiac murmur, and hemiplegia. Despite improved diagnosis, medical therapy, and surgical valve replacement, IE remains a disease with high morbidity and mortality in the 21st century. Neurologic complications are the most frequent extracardiac complications of left-sided IE, occurring in 20%—55% of patients. Because of the high incidence rate of neurological complications, the mortality and disability rate of patients with IE is significantly increased. Therefore, it is very important to identify neurological complications in patients with IE. Neurologic complications may impact diagnosis and therapeutic plans, particularly in patients requiring urgent cardiac surgery.

The neurological complications caused by IE can be divided into three types: cerebral infarction, intracerebral hemorrhage and CNS infection. Cerebral infarction is the most common complication, accounting for 50%—70% in all neurological complications, and the incidence rate is 20%—30% in patients with IE. Cerebral infarction caused by IE is usually symptomatic and can be complicated by postinfarction hemorrhage. Although embolic complications can be the presenting symptom of IE, these often occur during the first week of antibiotic therapy. Intracerebral hemorrhage, such as intracranial hematoma and subarachnoid hemorrhage (SAH), ranges from 2% to 17%. Other neurological complications such as infection include encephalitis, meningitis, and abscess. In rare cases, endocarditis can be secondary to spinal cord infarction or abscess, discitis, retinal ischemia, or ischemic of brain and peripheral nervous system. The main risk factors for neurologic events in severe IE patients include *Staphylococcus aureus*, large vegetation measured on echocardiography (>10 mm), mitral valve involvement, and nonneurologic embolic events. More than two-thirds of patients admitted to the ICU with neurologic manifestations of IE die or have residual neurological sequelae.

Most of the CNS complications caused by IE are due to embolism. Whether it is septic or nonseptic embolism, the vessel often involved is the middle cerebral artery, leading to ischemic stroke in the blood supply area. Septic emboli from vegetations are commonly considered as the main mechanism responsible for the observed brain lesions in patients with IE. It causes cerebral small vessel arteritis or bacterial aneurysm leading to hemorrhagic stroke, brain microabscesses, and small ischemic lesions. Spread to the meninges may cause encephalitis and meningitis. CNS primary vasculitis is thought to be triggered by infections, and secondary CNS vasculitis is most common associated with infections. Cerebral hemorrhage can be secondary to cerebral embolism. Hemorrhagic lesions may involve other complex mechanisms.

Bacterial aneurysms are associated with endovascular infectious embolism. Infective embolus causes necrosis of the inner layer of local vascular lumen, destruction of the outer layer and muscle layer structure, and formation of tumor-like bulge under the action of pressure. Bacterial aneurysms are common in intracranial vessels, especially at the distal end of arterial branches. The middle cerebral artery is the most common site of bacterial aneurysms. 40% of bacterial aneurysms were located at the distal end of the middle cerebral artery. Intracranial abscess is very rare, and its incidence is less than 1%. Intracranial microabscesses are often caused by multiple cerebral infarctions and are usually associated with *S. aureus* infection.

Plain CT scan of brain is often the initial imaging because it is easily available and sensitive for the detection of brain hemorrhages and large ischemic stroke. Ischemic lesions in IE are characterized by small, multiple, and disseminated infarctions. Intracranial hematoma shows high-density shadow on CT. The bleeding sites can be divided into brain parenchyma, subarachnoid space, subdural, and intraventricular. The bleeding after cerebral infarction is transformed into high and low mixed density. When the hemorrhage is located in the subarachnoid space, it shows the "groove edge sign" along the surface of the gyrus. Some patients had hemorrhagic transformation after cerebral infarction, which showed high-density shadow within the low-density lesions. Neuroimaging for all patients, even those without clinical neurologic symptoms, is now considered routine. MRI has been shown to be more sensitive and more specific than CT. The use of brain MRI in IE may improve prognostic assessment and aid in decision-making, especially for indication and timing of cardiac surgery.

## 1. Ischemic lesion

Ischemic stroke is the most common neurological complication of IE. Most patients show multiple lesions with various patterns. Infarcts can be infra- or supratentorial. DWI sequences can detect acute and small ischemic lesions. MRI also noted the predominance of lesions affecting cortical and border-zone arterial territories. Regarding the

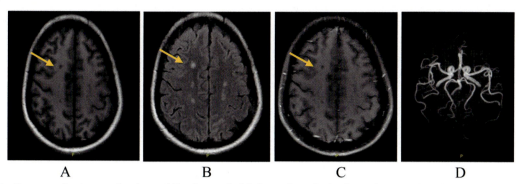

**FIGURE 5.2.1** Representative images showing multifocal watershed infarcts after mitral valve replacement for infective endocarditis. (A) and (B) Multiple speckled low-intensity on T1WI and hyper-intensity on FLAIR; (C): CE-T1WI showing no obvious abnormal enhancement in the above lesions; (D): TOF-MRA showing no obvious vascular stenosis and occlusion.

arterial territories, cortical branch infarction is the most common lesion, which usually involved the distal middle cerebral artery tree (Fig. 5.2.1).

## 2. Cerebral microbleeds (CMBs)

T2*-weighted sequences are used to detect small foci of hemorrhages. GRE sequences detect round or ovoid signal with a diameter<10 mm. These lesions have a strong specific association with IE than with other cerebral diseases. CMBs may represent pyogenic vasculitis or subacute inflammatory microvascular process. Hyperintense T2 halo and T1 enhancement is also a specific inflammatory characteristic of cerebral microbleeds (CMBs) associated with IE (Fig. 5.2.2).

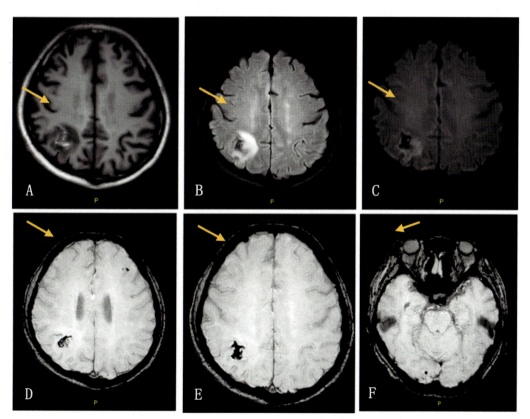

**FIGURE 5.2.2** MRI of a patient with infectious endocarditis and hemorrhagic transformation of ischemic lesions in right parietal lobe. (A) and (B): T1WI and FLAIR showing a mixed abnormal signal shadow in right parietal lobe with patchy edema intensity; (C): DWI showing limited diffusion of the above lesion; (D—F): Susceptibility-weighted imaging (SWI) detects microbleeds in the left frontal lobe and right occipital lobe.

## 3. Hemorrhagic lesions

Cerebral bleeding includes SAHs, lobar hemorrhages, and hemorrhagic transformation of an ischemic stroke. Acute SAH can be detected on CT and T2* sequences (Figs. 5.2.3 and 5.2.4).

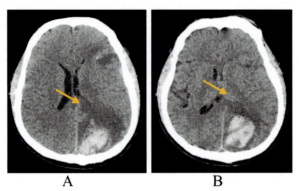

A  B

**FIGURE 5.2.3** (A) and (B). Axial brain CT images of a patient with multiple hemorrhages in left frontal and occipital lobes after mitral valve replacement for infective endocarditis, presenting as largely flaky homogeneous hyperdensity.

**FIGURE 5.2.4** Axial brain MRI images and TOF-MRA of an IE patient with hemorrhagic transformation of ischemic lesions in right frontal lobe. DWI and SWI sequences detected microbleeds in the lesion. (A) and (B): T1WI and FLAIR showing a mixed abnormal signal shadow in right frontal lobe with patchy edema intensity. (C): DWI showing limited diffusion of the above lesion. (D): Susceptibility-weighted imaging (SWI) detects microbleeds in left right frontal lobe. (E): TOF-MRA showing no obvious vascular stenosis and occlusion.

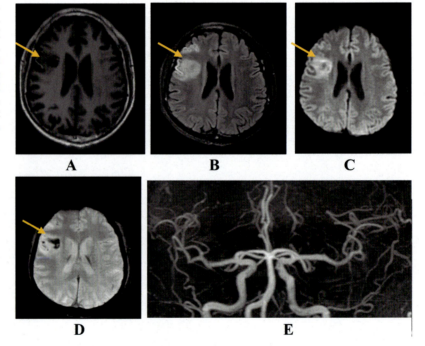

A  B  C

D  E

## 4. Intracranial mycotic aneurysms

Microbial aneurysms can be seen with time-of-flight—weighted imaging, angiography, and three-dimensional T1-weighted imaging after gadolinium enhancement. Mycotic aneurysms are usually multiple, bilateral, distal, and fusiform. The middle artery territory is usually involved. Conventional angiography is the best choice to ensure a complete and exact diagnosis. This technique allows aneurysm evaluation and treatment. Coiling may be required as a preventive measure in case of nonaneurysm involution under suitable antibiotic therapy.

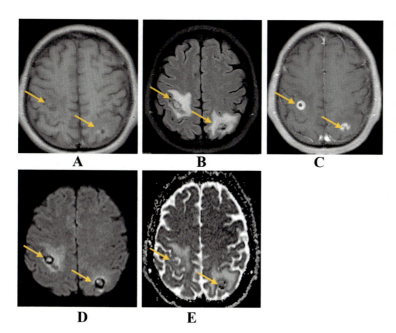

**FIGURE 5.2.5** Axial brain MRI images and DWI of an infective endocarditis patient with multiple microabscess in bilateral parietal lobe and ring enhancement. (A) and (B). T1WI and FLAIR showing mixed signal nodules in the bilateral parietal lobe with small patchy edema. (C). CE-T1WI showing ring enhancement of the above-mentioned lesions. (D) and (E). DWI and ADC map showing the central of the lesion was diffusion limited, and no diffusion limitation was found in the wall of the lesion.

## 5. Brain abscess and meningitis

A macroabscess is typically seen as an expansive lesion with central restricted diffusion, hyperintense peripheral edema on the FLAIR sequence, and postcontrast annular enhancement. Microabscesses are more frequent and smaller than 1 cm. It is not always straightforward to differentiate microabscesses and septic embolism, most particularly for small and pre-suppurative lesions. Meningeal irritation, lepto- and pathy-meningeal contrast enhancement as a nonspecific T2-FLAIR hyperintense signal of the subarachnoid is more readily seen on MRI than CT does (Fig. 5.2.5).

## Further reading

[1] Sila CA. Neurological complications of bacterial endocarditis. Handb Clin Neurol 2010;96:221−9.

[2] Champey J, Pavese P, Bouvaist H, Kastler A, Krainik A, Francois P. Value of brain MRI in infective endocarditis: a narrative literature review. Eur J Clin Microbiol Infect Dis February 2016;35(2):159−68.

[3] Cruz-Flores S. Neurologic complications of valvular heart disease. Handb Clin Neurol 2014;119:61−73.

[4] Cantier M, Mazighi M, Klein I, Desilles JP, Wolff M, Timsit JF, et al. Neurologic complications of infective endocarditis: recent findings. Curr Infect Dis Rep September 19, 2017;19(11):41.

[5] Colen TW, Gunn M, Cook E, Dubinsky T. Radiologic manifestations of extra-cardiac complications of infective endocarditis. Eur Radiol November 2008;18(11):2433−45.

[6] Sheth KN, Nourollahzadeh E. Neurologic complications of cardiac and vascular surgery. Handb Clin Neurol 2017;141:573−92.

[7] Dhar R. Neurologic complications of transplantation. Handb Clin Neurol 2017;141:545−72.

[8] Shen G, Shen X, Pu W, Zhang G, Lerner A, Gao B. Imaging of cerebrovascular complications of infection. Quant Imag Med Surg November 2018;8(10):1039−51.

Chapter 5.3

# Neurological complications of hypertension

**Teng Jin**

Both acute and chronic hypertension can cause neurological complications. The main complication of acute hypertension is stroke. Complications of chronic hypertension are mainly stroke, including cerebral infarction, intracranial hemorrhage, and subarachnoid space. Hypertension is defined as systolic blood pressure $\geq 140$ mmHg or diastolic blood pressure $\leq 90$ mmHg [1]. When blood pressure rises sharply, vasoconstriction is uneven, which easily leads to vascular endothelial damage and blood–brain barrier damage [2], thereby leading to hypertensive encephalopathy. Chronic hypertension can lead to intracranial vascular remodeling, intimal hyperplasia, and lumen narrowing [3], thereby reducing vascular wall tension.

## 1. Aneurysmal subarachnoid hemorrhage

Nontraumatic "spontaneous" SAH represents about 5% of all acute "strokes" [4]. The most common cause of nontraumatic SAH is a ruptured intracranial saccular ("berry") aneurysm (aSAH). Because most saccular aneurysms are located either on the circle of Willis or at the middle cerebral artery bifurcation, the most common locations for aSAH are the suprasellar cistern and sylvian fissures. Aneurysmal SAH can be focal or diffuse. Attempts to determine the precise anatomic location of a suspected intracranial aneurysm based on the distribution of SAH are not specific. Anterior interhemispheric aSAH is typically associated with rupture of a superiorly directed anterior communicating artery aneurysm.

   CT findings: (1) high-density shadow in the sulcus; (2) distribution varies with aneurysm location; (3) saccular aneurysm: Most commonly found at the circle of Willis or at the bifurcation of the middle cerebral artery; (4) CTA 90% –95% positive if aneurysm $\geq 2$ mm (Fig. 5.3.1).

## 2. Hypertensive intracranial hemorrhage

Intraparenchymal hemorrhage represents about 15% of all strokes and includes multiple etiologies. Hypertensive hemorrhage is the most common etiology [5], representing about 40%–60% of all primarily intracranial hemorrhages. Other etiologies include amyloid angiopathy in elderly patients, as well as vascular malformations, vasculitis, drugs, and bleeding diathesis. Risk factors for hemorrhagic stroke include increasing age, hypertension, smoking, excessive alcohol consumption, prior ischemic stroke, abnormal cholesterol, and anticoagulant medications. Although the MR physics related to hemorrhage are complex, the stages are generally accepted as hyperacute, acute, early subacute, late subacute, and chronic.

**FIGURE 5.3.1** (A). axial CT scan in a 43-year-old man shows multiple high-density in bilateral sylvian cisterns and annular cisterns, and the temporal horn of the lateral ventricle; (B). Image of the CTA obtained in the same patient shows a saccular aneurysm projecting superiorly from the left internal carotid artery (C7 segment).

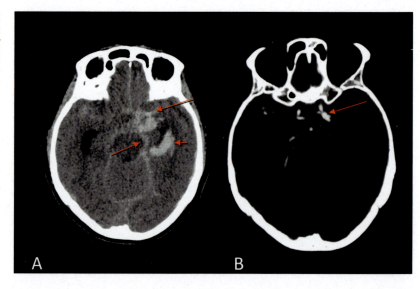

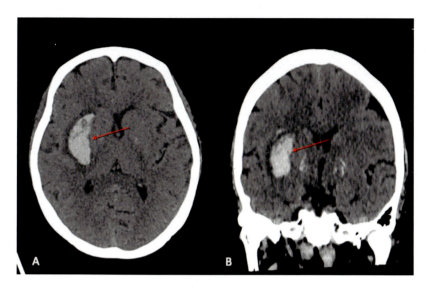

**FIGURE 5.3.2** CT scan of a 55-year-old hypertensive man shows hemorrhage in right basal ganglia and external capsule with surrounding edema and mass effect (see arrows in A and B).

CT findings: (1) Round or oval hyperdense mass in basal ganglia (BG) or thalamus in patients with hypertension; (2) Heterogeneous density if coagulopathy or active bleeding; (3) Intraventricular hemorrhage is common (Fig. 5.3.2).

## 3. Lacunar infarcts

Lacunar infarcts refer to infarcts <2 cm directly due to occlusion of small perforating arteries [6]. More than 50% of lacunar infarcts are located in BG, thalamus, and the rest are located in the internal capsule, pons, and subcortical white matter (Fig. 5.3.3). Hypertension is the most important risk factor for lacunar lesions. Lacunar cerebral infarction has some

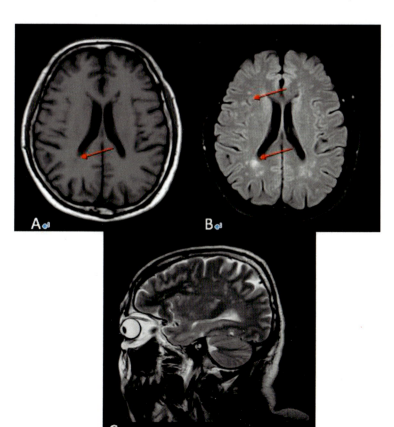

**FIGURE 5.3.3** A 60-year-old female with axial MR images shows bilateral periventricular hypointense foci in T1WI (arrows A) and hyperintensity in T2-FLAIR (arrows B) indicating chronic ischemic changes.

classic clinical manifestations; the most common is pure motor hemiplegia, accounting for 45%. Movement disorders may involve the face, upper extremities, and lower extremities.

Imaging: Small (commonly 3−15 mm), well-circumscribed areas of parenchymal; abnormality (encephalomalacia) in BG, thalamus, WM. MRI: T1WI, Small, well-circumscribed, hypointense foci; T2WI: Small, well-circumscribed, hyperintense foci; FLAIR: Increased signal; DWI: Restricted diffusion (hyperintense) if acute/subacute.

## 4. Binswanger syndrome

In 1994, Binswanger described a type of early dementia distinct from syphilitic paralytic dementia and macrovascular-associated vascular dementia, which can involve subcortical structures [7]. Hypertension is an important risk factor, present in 94% of individuals with dementia. Pathological changes were multiple lacunar infarcts with diffuse loss of myelin sheath. The periventricular area is particularly prominent. Patients experience progressive decline in motor, cognitive, and behavioral abilities over 5−10 years. The imaging findings showed hyperintensity in periventricular white matter on T2WI, overlapping with local subcortical lacunar infarcts (Fig. 5.3.4). Taking aspirin and controlling high blood pressure can slow the progression of the disease.

**FIGURE 5.3.4** A 65-year-old male with a long history of chronic hypertension. (A−C): T2 FLAIR images show symmetrical hypo and hyperintensity in bilateral periventricular, deep white matter, bilateral basal ganglia, and external capsules. This disease generally does not involve bilateral temporal lobes. D: No obvious enhancement of the lesions was noted.

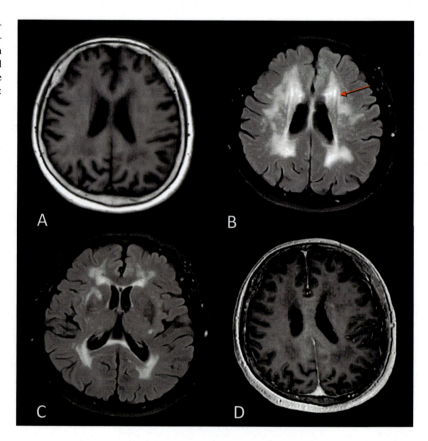

# References

[1] Finocchi C, Sassos D. Headache and arterial hypertension. Neurol Sci 2017;38(Suppl. 1):67−72.

[2] Lu Y, chen R, Cai J, et al. The management of hypertension in women planning for pregnancy. Br Med Bull 2018;128(1):75−84.

[3] Rivera SL, Martin J, Landry J. Acute and chronic hypertension: what clinicians need to know for diagnosis and management systemic vasculitis and nervous system. Crit Care Nurs Clin 2019;31(1):97−108.

[4] Heinz R, Brandenburg S, Nieminen-Kelha M, et al. Microglia as target for anti-inflammatory approaches to prevent secondary brain injury after subarachnoid hemorrhage (SAH). J Neuroinflammation 2021;18(1):36.

[5] Gross BA, Jankowitz BT, et al. Cerebral intraparenchymal hemorrhage: a review. JAMA 2019;321(13):1295−303.

[6] Chojak-Lukasiewicz J, Dziadkoiak E, et al. Cerebral small vessel disease: a review. Adv Clin Exp Med 2021;30(3):349−56.

[7] Rosenberg GA. Binswanger's disease: biomarkers in the inflammatory form of vascular cognitive impairment and dementia. J Neurochem 2018;144(5):634−43.

Chapter 5.4

# Neurological complications of systemic vasculitis

**Teng Jin**

Systemic vasculitis is a group of diseases with a high degree of clinical heterogeneity, with inflammation of the vessel wall as the main manifestation [1]. At least 20 types of systemic vasculitis have been recognized so far. Nervous system damage is one of the most common clinical manifestations of systemic vasculitis, which can involve the central and peripheral nervous systems [2]. Primary vasculitis: Involvement of large vessels: (1) Takayasu arteritis (TA); (2) Giant cell arteritis (GCA); (3) Behcet's disease; Involvement of middle vessels: (1) Arteritis nodosa; (2) Thromboangiitis obliterans (Buerger's disease); Involvement of small vessels: (1) Weger's granulomatosis; (2) Churg—Strauss syndrome.

## 1. Bechet's disease

Bechet's disease (BD) is a systemic, chronic, vascular inflammatory disease. The main clinical manifestations include recurrent oral ulcers, genital ulcers, ophthalmia, and skin lesions. It can also involve the nervous system, digestive tract, and joints [3]. The current general diagnostic performance was developed by the International BD Research Group in 1900, and the standard has sensitivity of 91% and specificity of 96%. The disease can occur at any age, but the most common age is 16—40 years old [4]. In China, it is more common in women, and the vascular, nervous system, and eye involvement in male patients is more serious than that in female patients (Fig. 5.4.1).

**FIGURE 5.4.1** A 55-year-old female patient with Bechet's disease. (A—C): T1 T2 FLAIR images showing hypo- and hyperintensity in the left basal ganglia region with tumor-like contrast enhancement. (D): MRS shows the reduction of NAA peak to varying degrees.

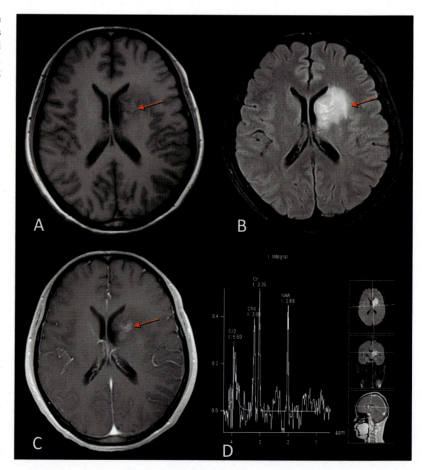

## 2. Takayasu arteritis

TA refers to a chronic progressive, nonspecific inflammatory disease of the aorta and its major branches. The lesions are common in the aortic arch and its branches, followed by the descending aorta, abdominal aorta, and renal artery [5]. Secondary branches of the aorta, such as the pulmonary artery and coronary arteries, can also be affected. The affected vessel can be full-thickness arteritis. In the early stage, the lining of blood vessels was infiltrated by lymphocytes, and plasma cells. Polymorphonuclear neutrophils and multinucleated giant cells can be seen occasionally. Due to the thickening of the intima of the blood vessel, the stenosis or occlusion of the vessel is often found. In a small number of patients, the medial layer, elastic fibers, and smooth muscle fibers of the arterial wall are damaged due to inflammation, resulting in arterial dilatation and pseudoaneurysm [6]. Imaging: TA presents with thickening of the walls of large vessels, usually involving the thoracic aorta and its branches, mostly the left subclavian artery. Imaging typically presents with segmental stenosis, occlusion, or aneurysm of the aorta and branches (Fig. 5.4.2).

## 3. Giant cell arteritis

The cause of GCA is unknown. The greatest risk factor for developing GCA is advanced age. The disease never occurs before age of 50, and its incidence steadily increases after age 50. Ethnicity and geography are also important risk factors, with the highest incidence rates seen in Scandinavia and among people of Scandinavian immigrant descent in the United States [7]. There is also a genetic susceptibility to the occurrence of GCA, and recent studies have confirmed that GCA is associated with genes in the human leukocyte antigen class II region. The investigation of family incidence showed that the first-degree relatives of GCA patients were more affected, and most of them had HLA-DR4 and CW3 [8], suggesting a genetic susceptibility. In addition, there are environmental risk factors for the onset of GCA, as well as gender and healthy status. GCA inflammatory response is concentrated in elastic lamina of arteries, which may be related to some of its self-antigens. Immunohistochemical studies also found immunoglobulin deposits in the inflamed temporal artery wall, infiltrating inflammatory cells are mainly TH cells, and lymphocytes in the patient's peripheral blood are sensitive to human arteries and myokines in vitro (Fig. 5.4.3).

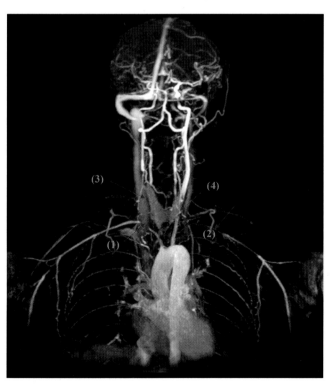

**FIGURE 5.4.2**  Severe stenosis/occlusion in brachiocephalic trunk and proximal right subclavian artery. Segmental occlusion in left subclavian artery. Complete occlusion in right common carotid artery; Moderate to severe stenosis in proximal left common carotid artery.

**FIGURE 5.4.3** A 27-year-old female with giant cell arteritis. The walls of the aortic arch and descending aorta are unevenly thickened and the vessel is dilated.

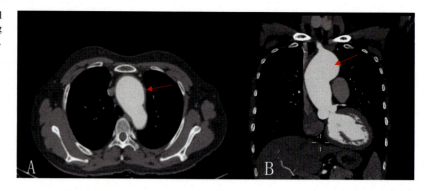

## 4. Wegner's granulomatosis

Wegner's granulomatosis is a necrotizing granulomatous vasculitis, which belongs to autoimmune disease. The lesions involve small arteries, veins, and capillaries, and occasionally large arteries. The pathology is characterized by inflammation of the blood vessel wall. Features mainly involve the upper and lower respiratory tract and kidneys [9]. It usually begins with focal granulomatous inflammation of the nasal mucosa and lung tissue and progresses to diffuse necrotizing granulomatous inflammation of the blood vessels. Clinical symptoms usually manifest as inflammation in nasal, pulmonary, and renal tissue, and joints. Early treatment can be done with combination of glucocorticoids and cyclophosphamide.

## References

[1] Younger DS, Carlson A. Dermatologic aspects of systemic vasculitis. Neurol Clin 2019;37(2):465−73.

[2] Richard Kitching A, Anders H-J, Basu N. ANCA-associated vasculitis. Nat Rev Dis Prim 2020;6(1):71.

[3] kanakis MA, Vaiopoulos AG, Vaiopo GA, et al. Epididymo-orchitis in bechet's disease: a review of the wide spectrum of the disease. Acta Med Iran 2017;55(8):482−5.

[4] Kumar A, Sahu A, Kaushik J, et al. Kyrieleis like plaques - atypical presentation of ocular Behcet's disease. Roman J Ophthalmol 2021;65(4):383−5.

[5] Tombetti E, Mason JC. akayasu arteritis: advanced understanding is leading to new horizons. Rheumatology 2019;58(2):206−19.

[6] Esatoglu SN, Hatemi G. Takayasu arteritis 2022;34(1):18−24.

[7] Ciofalo A, Gulotta G, Iannella G, et al. Giant cell arteritis (GCA): pathogenesis, clinical aspects and treatment approaches. Curr Rheumatol Rev 2019;15(4):259−68.

[8] Lyons HS, Quick V, Sinclair AJ, et al. A new era for giant cell arteritis. Eye 2020;34(6):1013−26.

[9] Ahmadia H, Haddad AH, Darabi Mahboub MR. A nonspecific penile ulcer leading to the diagnosis of wegener's granulomatosis. Urol J 2020;17(2):210−2.

# Chapter 6

# Digestive system

Peng Liu

*Department of Radiology, Hunan Provincial People's Hospital, First Affiliated Hospital of Hunan Normal University, Changsha, Hunan, China*

## 1. Introduction

The digestive system (Latin: systema digestorium), also known as alimentary system, is digestive tract from the mouth to anus, with its associated glands and organs. The function of the digestive tract is mainly to provide mechanical and chemical food procession, absorption of nutrients, and excretion of undigested remains. In addition to secreting bile to participate in the digestion of lipids, the liver is also an important site of substance metabolism. The pancreas has exocrine functions and is an endocrine organ involved in blood glucose regulation. A variety of conditions can affect both nervous system and digestive system. These diseases are mainly divided into three categories. The first category is neurological changes caused by abnormal absorption or metabolism of nutrients, including hepatic encephalopathy (HE), pancreatic encephalopathy, hepatic myelopathy (HM), Wernicke's encephalopathy, and inflammatory bowel disease (IBD) involving the central nervous system (CNS). The second category is the manifestation of congenital syndromes in the digestive and nervous systems, including Wilson disease, tuberous sclerosis (TS), multiple endocrine neoplasia type 1 (MEN1), von Hippel-Lindau (VHL) syndrome, hereditary hemorrhagic telangiectasia (HHT), Turcot syndrome, neurofibromatosis type 1 (NF1), Gardner syndrome, etc. The third category is systemic diseases involving both the digestive and nervous systems, including IgG4-related disease, sarcoidosis, Behçet disease, systemic sclerosis, amyloidosis, etc.

## 2. Hepatic encephalopathy

HE, also known as portosystemic encephalopathy, refers to a spectrum of neuropsychiatric abnormalities in liver dysfunction and portal hypertension patients. Excessive amounts of ammonia from the digestive system or elsewhere, which is usually metabolized by the liver reaches the systemic circulation, are taken up by the brain and cause toxic effects on astrocytes and neurons [1]. The vast majority of patients have portosystemic shunts in the setting of cirrhosis, either from the development of spontaneous shunting or as a result of transjugular intrahepatic portosystemic shunting. Clinical manifestations of HE range widely from chronic episodic subclinical neurological dysfunction to acute fulminant neurological impairment, coma, and death [2]. A classically described manifestation of HE is asterixis or flapping tremor, which appears in intermediate phases. Its presence indicates an overt HE. In HE, neurological abnormalities occur after an established liver disease.

Radiographic features include acute hyperammonemia and chronic liver disease. In mild cases, symmetric hyper-intensity within the insula (most common), thalamus, posterior limbs of internal capsule, and cingulate gyrus is seen on T2/FLAIR images. This abnormal signal is often reversible with therapy. In severe cases, there are diffuse cortical edema and hyperintensity on T2/FLAIR images. The perirolandic and occipital regions are typically spared. MR spectroscopy may show elevated glutamine/glutamate peak coupled with decreased myoinositol and choline [3,4]. Imaging finding of chronic liver disease is mainly cirrhosis, which includes surface and parenchymal nodularity, parenchymal heterogeneity, and signs of portal hypertension.

## 3. Acquired hepatocerebral degeneration

Acquired hepatocerebral degeneration is an uncommon irreversible extrapyramidal neurodegenerative condition encountered in patients with cirrhotic chronic liver disease, resulting in widespread cerebral, basal ganglia, and cerebellar damage.

**FIGURE 6.1** A 46-year-old woman with cirrhosis. (A) Brain MRI images demonstrate T1-weighted hyperintensity within bilateral globus pallidus (*arrows*). (B) Abdominal CT images demonstrate liver surface nodularity and lower esophageal varices (*arrow*).

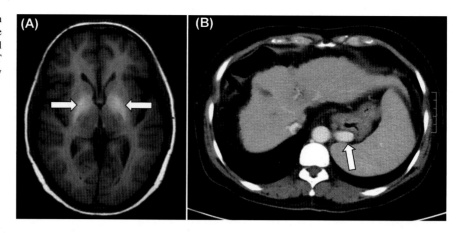

Acquired hepatocerebral degeneration is a term that is restricted to patients which cirrhotic liver disease resulting from a variety of causes but specifically excluding Wilson disease. Patients with acquired hepatocerebral degeneration usually, but not always, have had multiple prior episodes of HE [5]. They present with gradual neurological dysfunction, including dementia, rigidity, dysarthria, gait ataxia, tremor, and choreoathetosis [6]. Although the pathophysiology of acquired hepatocerebral degeneration is uncertain, manganese overload is believed to be part of the disease and responsible for T1 shortening observed in globus pallidus [7]. In acquired hepatocerebral degeneration, neurological abnormalities occur after an established liver disease.

Radiographic features include manganese overload in the brain and chronic liver disease. Intrinsic high signal intensity in globus pallidus ± subthalamic region (Fig. 6.1) and midbrain is seen on T1-weighted image. This abnormal signal may reverse following liver transplantation. T2-weighted image may show increased signal in the middle cerebellar peduncles [6]. The liver imaging is similar to the findings of HE.

## 4. Pancreatic encephalopathy

Pancreatic encephalopathy is the occurrence of neuropsychiatric abnormalities in acute pancreatitis (AP), which is not otherwise explained by the presence of electrolyte abnormities or organic lesions. The pathogenesis of pancreatic encephalopathy is incompletely understood. The mechanism recognized by most scholars is brain tissue damage caused by abnormal enzyme metabolism based on severe pancreatitis [8,9]. It is classified into pancreatic encephalopathy of early stage (usually 2−5 days after AP) and late stage (after 2 weeks or during convalescence) [10]. Early stage is considered a part of septic encephalopathy, and the late stage may result from the loss of vitamin B1. Clinically, it is similar to other encephalopathies. There is no correlation between the severity of pancreatitis and incidence of this condition [8]. It usually presents early in the disease and shows many neuropsychiatric manifestations, including altered sensorium, confusion, agitation, seizures, speech disorders, and hallucinations [11]. Short intervals of lucid period are interspersed in these neuropsychiatric phases. There is a cyclic progression with remission and relapses [12].

In pancreatic encephalopathy, imaging abnormalities occur after established AP. Patchy white matter lesions are known in the condition (Fig. 6.2), but neuroimaging may be normal [11]. These patchy signal abnormalities may be seen on MRI in cerebral white matter and resemble multiple sclerosis. Imaging of AP includes focal or diffuse parenchymal enlargement, indistinct pancreatic margins, surrounding retroperitoneal fat stranding, and liquefactive necrosis of parenchyma.

## 5. Hepatic myelopathy

HM is a rare neurological complication of chronic liver disease, which is usually seen in adults, presenting as pure motor spastic paraparesis. It is almost always associated with portosystemic shunts and hepatic decompensation. Symptoms typically worsen during HE episodes. The pathogenesis is unclear, mainly due to the combined effects of chronic poisoning, nutritional deficiency, immune damage, and hemodynamic changes [13−16]. The pathological changes include demyelination of the spinal cord below cervical spinal cord and the entire length of spinal cord, bilateral symmetrical demyelination with degeneration and disappearance of nerve axons, and decrease of nerve fibers and replacement by glial cells [17,18]. In HM, neurological abnormalities occur after established liver disease.

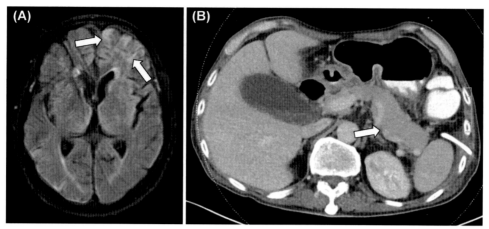

**FIGURE 6.2** A 58-year-old man, during convalescence of severe acute pancreatitis, presented with somnolence and delirium. (A) Brain MRI demonstrates T2/FLAIR hyperintensity within subcortical white matter (*arrows*). (B) Abdominal CT demonstrates mild swelling of pancreas (*arrow*).

Abdominal ultrasonography primarily indicates liver cirrhosis (portal vein inner diameter>13 mm). Negative spinal cord MRI results supported HM in the differential diagnosis because MRI was essential to rule out such etiologies involving infarction, myelitis, or compression of spinal cord (epidural cord compression or intrinsic cord tumor). Brain MRI showed hyperintensity on T1WI within basal ganglia, including pallidum, lentiform nucleus, putamen, internal capsule, and extensive white matter disease on T2WI [19,20]. The toxic effect of portal blood is most significant in the basal ganglia, probably because of high metabolic activity in this area.

## 6. Wernicke encephalopathy

Wernicke encephalopathy, also referred as Wernicke–Korsakoff syndrome, is a form of thiamine (vitamin B1) deficiency and is typically seen in alcoholics. It was originally described as characterized by triad of acute confusion, ataxia, and ophthalmoplegia (most commonly horizontal nystagmus and conjugate gaze palsies). Thiamine deficiency results from malnutrition or malabsorption, which can occur for many reasons (typically prolonged total parenteral nutrition without supplementation or postbariatric surgery) [21]. In Wernicke encephalopathy, neurological abnormalities occur after potential digestive system disease.

It is commonly seen on MRI as areas of symmetrical increased T2/FLAIR signal involving mammillary bodies, dorsomedial thalami, tectal plate, periaqueductal area, and/or around third ventricle (Fig. 6.3). Contrast enhancement and restricted diffusion can also be seen in the same regions, most commonly mammillary bodies [22]. MR spectroscopy may show decreased or normal N-Acetyl Aspartate (NAA) with notable presence of lactate. Abdominal imaging may indicate the cause of Wernicke encephalopathy, such as intestinal fistula, AP, etc.

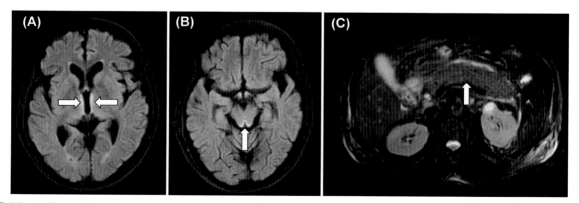

**FIGURE 6.3** A 66-year-old man with wernicke encephalopathy. He underwent prolonged total parenteral nutrition due to acute pancreatitis. (A, B) Brain MRI demonstrates T2-FLAIR hyperintensity within bilateral dorsomedial thalami and periaqueductal area (*arrows*). (C) Abdominal MRI demonstrates marked swelling of pancreas (*arrow*), consistent with acute pancreatitis.

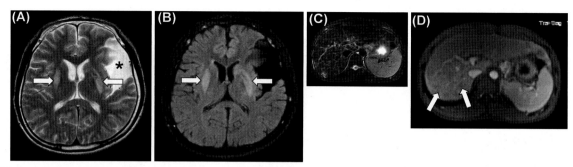

**FIGURE 6.4**  A 20-year-old man with wilson disease. (A, B) Brain MRI demonstrates T2/FLAIR hyperintensity within bilateral putamina (*arrows*); arachnoid cyst in the left middle cranial fossa is noted (*asterisk*). (C, D) Liver MRI demonstrates cirrhotic change and diffuse regenerative nodules (*arrows*).

## 7. Inflammatory bowel disease CNS involvement

IBD is a chronic remitting and relapsing disorder of the bowel, usually presenting as ulcerative colitis or Crohn's disease. CNS involvement belongs to extraintestinal manifestations of IBD. Cerebrovascular disease, the most commonly reported neurological complication of IBD, has been attributed to hypercoagulability, vasculitis, or consumption coagulopathy [23−25]. Neuropsychiatric manifestations include stress-evoked responses, emotional disturbances, depression, and impaired cognitive functioning. It is commonly seen on T2/FLAIR as focal white matter hyperintensities [26]. Different types of bowel wall thickening and its complication are seen on abdominal study depending on period of the disease.

## 8. Wilson disease

Wilson disease, also known as hepatolenticular degeneration, is a multisystem disease due to abnormal accumulation of copper caused by a variety of mutations in *ATP7B* gene. It is characterized by early-onset liver cirrhosis, with CNS findings most frequently affecting basal ganglia and midbrain. Common neurological clinical features include dysarthria, dystonia, tremor, parkinsonism, choreoathetosis, and ataxia and gait anomalies. The form of liver disease varies depending on the severity of the disease at the time of diagnosis and pathological findings include fatty changes, acute hepatitis, chronic active hepatitis, cirrhosis, and occasionally fulminant hepatic necrosis. Asymptomatic Kayser−Fleischer rings are usually seen in the cornea and are a characteristic but nonpathognomonic feature [4,27].

Neuroimaging features of Wilson disease may vary depending on whether the disease is treated or untreated. Basal ganglia (especially putamen) is the most frequently affected site, followed by midbrain, pons, and thalamus [28,29]. The distribution is bilateral and symmetric [30]. Abnormal T2 hyperintensity in putamina is the most common MRI abnormality (Fig. 6.4). Additional areas of T2 signal abnormality predominantly affect deep gray nuclei. Involvement of midbrain tegmentum can appear as a face of giant panda sign on axial images. Axial T2WI at pons may also show the face of a miniature panda sign (cub of the giant panda). This combination is referred to as double panda sign [31]. On T1WI in patients presenting with neurologic manifestations, these areas show hypointensity. In contrast, patients with severe hepatic dysfunction show areas of T1 hyperintensity, especially in globus pallidus, similar to that seen in acquired (non-Wilsonian) hepatocerebral degeneration attributed to manganese deposition. At CT, hepatic attenuation can be increased [32] or normal, where the latter is thought to result from fatty infiltration and copper deposition canceling effects of each other. MRI demonstrates the contour abnormalities and parenchymal nodules of liver in more than half of patients with Wilson disease, which correlates with the severity of hepatic dysfunction and clinical manifestations.

## 9. Tuberous sclerosis

TS, also known as TS complex or Bourneville disease, is a neurocutaneous syndrome characterized by the formation of nonmalignant hamartomas and neoplastic lesions in the brain, heart, skin, kidney, lung, and other organs. It was classically described as presenting in childhood with a pathognomonic triad (Vogt triad) of seizures, intellectual disability, and adenoma sebaceum [33]. Spontaneous mutations account for 50%−86% of cases [34], with the remainder inherited as an autosomal dominant condition.

The common neuroimaging features include cortical/subcortical tubers, subependymal hamartomas, subependymal giant cell astrocytomas, and white matter abnormalities (Fig. 6.5). Cortical or subependymal tubers frequently calcify after

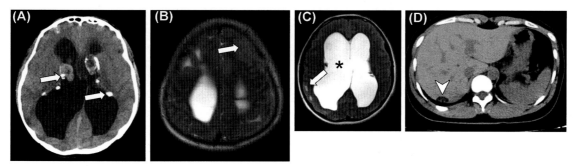

**FIGURE 6.5** A 23-year-old man with tuberous sclerosis. (A) Brain CT demonstrates multiple subependymal hamartomas with calcification (*arrows*). (B, C) Brain MRI demonstrates T2-weighted hyperintensity within the subcortical white matter (*arrows*). Note the prominent hydrocephalus (*asterisk*). (D) Abdominal CT demonstrates a lobulated fat-containing lesion in right hepatic lobe (*arrowhead*), consistent with angiomyolipoma.

2 years of age. The linear bands radiating from periventricular white matter to subcortical region are thought to be specific for TS [35]. The digestive system manifestation of TS is hepatic angiomyolipoma (AMLs). Only 6% of hepatic AMLs are associated with TS, compared to up to 20% of renal AMLs [36]. AMLs may be single or multiple, round or lobulated fat-containing mass lesions, seen more commonly in right hepatic lobe [37]. Due to the presence of vascular component, marked enhancement in arterial phase is evident. Drainage is via hepatic veins, and this is the main differentiating point from fat-containing Hepatocellular carcinoma (HCC) drains mainly in portal vein.

## 10. Multiple endocrine neoplasia type 1

MEN1, also known as Wermer syndrome, is an autosomal dominant genetic disease resulting in proliferative lesions in multiple endocrine organs, particularly the pituitary gland, islet cells of pancreas, and parathyroid glands. The abnormality is related to *MEN1*, a tumor suppressor gene located on chromosome 11q13, which produces menin, a nuclear protein important for the regulation of gene expression. In addition to the characteristic lesions involving pituitary, parathyroid, and pancreas, numerous other lesions are encountered with greater frequency in patients with MEN1. These include carcinoid tumors, hepatic focal nodular hyperplasia [38], meningiomas [39], etc. Pancreatic islet cell tumor with associated hypersecretion syndromes is the most common presentation except for primary hyperparathyroidism. Gastrinomas are most common and associated with Zollinger—Ellison syndrome, followed by glucagonoma [40].

Most adenomas are hypointense relative to normal pituitary gland at cross-sectional imaging, both prior to and following intravenous administration of contrast material. The majority of pancreas and duodenum neuroendocrine tumors in MEN1 are small (<2 cm) and may be multiple; a larger size and the presence of calcification suggest malignancy. The tumors are typically iso-attenuating at unenhanced CT and enhance avidly in arterial phase.

## 11. Von Hippel-Lindau disease

VHL disease is a rare, inherited, multisystem disorder that is characterized by the development of numerous benign and malignant tumors in different organs (at least 40 types) due to mutations in VHL tumor suppressor gene on chromosome [41,42]. Clinical presentation is varied, depending on sites of disease. Most commonly, these are either within the abdominal cavity or affect the CNS. Pancreatic lesion may be the earliest manifestation [42], which includes cysts (50%—91%) (Fig. 6.6), pancreatic neuroendocrine tumors (pNET) (5%—17%), serous microcystic adenomas (12%), and rarely adenocarcinoma. pNET is usually nonfunctional [43] and frequently multiple. CNS manifestation includes hemangioblastoma(s) and choroid plexus papilloma. The most common location of CNS hemangioblastoma is cerebellar (60%), followed by spinal cord (30%). Spinal cord hemangioblastoma is commonly seen in cervical and thoracic cord. Only 5%—30% of all cerebellar hemangioblastomas are attributed to VHL disease, whereas 80% of spinal cord hemangioblastomas occur with the disease [44]. When associated with VHL, they occur at a younger age and have a worse prognosis. Symptoms of cerebellar lesions include headache, vertigo, ataxia, vomiting, nystagmus, and ninth cranial nerve palsy. Focal spinal pain is the most common symptom of hemangioblastoma [45].

Hemangioblastomas are highly vascular lesions that readily enhance with contrast medium. Spinal cord lesions may be associated with a syrinx. Pre- and postcontrast T1WI are the most useful sequence for detection of these lesions. Large feeding or draining vessels within periphery and a solid component may appear as tubular areas of flow void.

**FIGURE 6.6** A 48-year-old man with Von Hippel-Lindau disease. (A, B) CT images of abdomen with intravenous contrast material showed multiple pancreatic cysts (*arrows*) and a single spinal cord hemangioblastoma (*arrowhead*).

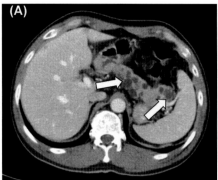
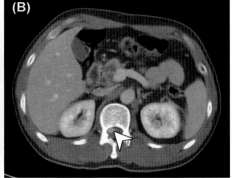

## 12. Hereditary hemorrhagic telangiectasia

HHT, also known as Osler—Weber—Rendu syndrome, is a rare inherited disorder characterized by abnormal blood vessel formation in the skin, mucous membranes, and organs, including the lungs, liver, and CNS. Classic clinical triad at presentation is epistaxis, multiple telangiectasias, and positive family history. It is an autosomal dominant multiorgan vascular dysplasia characterized by multiple arteriovenous malformations (AVMs) that lack an intervening capillary network. Telangiectasias (small superficial AVMs) are particularly common. Symptomatic liver involvement in HHT is uncommon but does occur [46,47]. It could cause high-output cardiac failure (arteriovenous or portovenous shunts), portal hypertension (arterioportal shunt), and biliary disease (arteriovenous or arterioportal shunting). AVMs or angiodysplasia is also seen in stomach, small bowel, or large bowel, complicated by recurrent gastrointestinal (GI) bleeding. CNS involvement in HHT includes cerebral AVMs, spinal AVMs, or cerebral aneurysms. One-third of cerebral complications in HHT are due to cerebral AVMs or aneurysms, and two-thirds are due to paradoxical emboli from pulmonary AVMs.

## 13. Turcot syndrome

Turcot syndrome is a historical term used to describe the association of primary CNS tumors and two different forms of colorectal polyps: familial adenomatous polyposis (FAP) and hereditary non—polyposis-related colorectal cancer (HNPCC) (or Lynch syndrome). It is characterized by multiple colonic polyps and an increased risk of colon and primary brain tumors. Turcot syndrome is a rare disease. Patients typically present in the second decade. It is thought to carry an autosomal recessive inheritance. Two-thirds of patients have mutations in APC gene, the same genetic defect as in FAP. These patients have multiple colonic adenomas, and virtually all develop colorectal carcinoma by the age of 40. The common intracranial tumor in this subtype is medulloblastoma. The other third has mutations in HNPCC genes. Colonic malignancy is not as common in this type but tends to develop at younger age. Most develop glioblastomas. MRI can be used to characterize CNS tumors seen in FAP and HNPCC. The anatomic location of medulloblastomas at MRI is different in molecular subtypes and age groups. There are no distinguishing features that are identified in imaging presentation of gliomas from sporadic types.

## 14. Neurofibromatosis type 1

NF1, also known as von Recklinghausen disease, is an autosomal dominant neurocutaneous disorder defined by the presence of café au lait spots, axillary freckling, plexiform neurofibromas, skeletal dysplasias, and iris hamartomas (Lisch nodules) [48]. Genetic mutation of the NF1 leads to abnormal tumor suppression, leading to a variety of benign and malignant tumors. NF1 confers at least a fivefold increased risk for developing gliomas, including World Health Organization (WHO) grade IV astrocytomas (such as glioblastomas). Glioblastomas usually occur in young adults, in whom the overall prognosis is poor. About 15%—20% of individuals with NF1 will develop low-grade (WHO grade I) glial neoplasms, 80% of which are in optic pathway and about 15% are in brainstem. Plexiform neurofibromas deform sphenoid wing and occipital bone, with associated dysplasia and proptosis best appreciated at radiography and CT. Benign neurofibromas are by far the most common tumor involving GI tract in NF1 patients. Gastrointestinal stromal tumors (GISTs), leiomyomas, leiomyosarcomas, carcinoids, and ganglioneuromas occur within the GI tract. Multiplicity is a common feature in the GI tract neurofibromas and GISTs. GISTs in NF1 patients are morphologically and immunohistochemically identical to tumors that occur in non-NF tumors. GISTs can be malignant, benign, or uncertain malignant potential whether they occur in NF1 or non-NF1 patients. Unlike neurofibromas, GISTs occur exclusively within small bowel. Intratumoral hemorrhage, necrosis, and degeneration can be seen.

## 15. Gardner syndrome

Gardner syndrome is one of the polyposis syndromes. It is characterized by FAP, multiple osteomas (especially of mandible, skull, and long bones), epidermal cysts, fibromatoses, desmoid tumors of mesentery, and anterior abdominal wall. There is an autosomal dominant inheritance in FAP gene (chromosome 5q) in most patients, but with 20% of cases resulting from new mutations. Extracolonic abdominal manifestations often precede the discovery of colonic polyps and include mesenteric and abdominal wall desmoid tumors, pancreatic neoplasms, hepatoblastoma, and hepatocellular carcinoma [49].

## 16. IgG4-related diseases

IgG4-related disease is a systemic disease characterized by fibroinflammatory infiltration of various organs, including plasma cells that express IgG4 (immunoglobulin G subclass 4). Immunohistochemistry demonstrate plasma cells that stain for IgG4. IgG4$^+$ plasma cells must account for more than 40% of IgG$^+$ plasma cells [50]. The absolute density of IgG4$^+$ plasma cells (number per high power field) suggestive of the diagnosis varies by organ. CNS manifestation of IgG4-RD includes IgG4-related hypertrophic pachymeningitis, IgG4-related hypophysitis, and chronic subdural hemorrhage. The most common affected digestive systems in IgG4-RD are the pancreas (autoimmune pancreatitis), bile ducts (sclerosing cholangitis), and salivary gland (IgG4-related sialadenitis). Diffuse enlargement of pancreas with loss of definition of pancreatic clefts in autoimmune pancreatitis results in "sausage-shaped pancreas." Focal disease is characterized by enlargement of pancreatic head or, less frequently, pancreatic body or tail, resulting in mass-like appearance. Intrahepatic and extrahepatic segments can be affected in sclerosing cholangitis, producing dense bile duct infiltration by IgG4-positive plasma cells and extensive fibrosis. The result is focal or diffuse thickening of bile duct wall, mostly associated with stenosis and upstream dilatation; findings are readily depicted on Endoscopic retrograde cholangiopancreatography (ERCP) or MR cholangiopancreatography [51]. ERCP usually showing multifocal and short intrahepatic biliary strictures with beaded or "pruned-tree" lesions is characteristic of primary sclerosing cholangitis. In contrast, long and continuous strictures are typically seen in patients with IgG4-related sclerosing cholangitis. In IgG4-related sialadenitis, patients present with either enlargement of lacrimal and salivary glands (an entity previously known as Mikulicz disease) or chronic sclerosing sialadenitis of submandibular glands (also known as Küttner tumor). Lesions in Mikulicz disease usually demonstrate homogeneous attenuation and enhancement on CT. On MRI, they have relatively hypo-intensity on T2WI owing to fibrosis and hypo-intensity with homogeneous enhancement on T1WI [52].

## 17. Sarcoidosis

Sarcoidosis is a multisystem disorder of unknown etiology characterized by formation of inflammatory noncaseating granulomas within affected tissues. It is thought to represent a disorder of immune regulation, particularly of cell-mediated immunity. Clinical presentation is variable, and diagnosis is usually made combining clinical and radiological features. 90% of patients have pulmonary involvement (although many are asymptomatic). Lacrimal gland and salivary gland involvement are relatively common. Liver is the most frequently involved viscera, with hypoattenuating/hypointense nodules ranging in size from 5 to 20 mm corresponding with coalescing granulomas [53,54]. The nodules become more confluent with increasing size.

On contrast-enhanced CT/MRI, liver nodules are hypo-enhancing relative to background parenchyma. Sarcoid can involve intra- or extrahepatic biliary tree. Intrahepatic involvement is granulomatous and produces cholestatic and primary biliary cirrhosis. Extrahepatic involvement results in stricture and cholestasis, which appears like cholangiocarcinoma. Stomach is the most common GI site of involvement. The radiological signs of the disease are nonspecific, ranging from mucosal thickening (mimicking Menetrier disease) to lesions mimicking gastric ulcers or linitis plastica [53]. Approximately 5% of patients develop neurosarcoidosis. Radiographic features of neurosarcoidosis can be thought of as occurring in one or more of five compartments. From superficial to deep, of which are skull vault involvement, pachymeningeal or leptomeningeal involvement, pituitary and hypothalamic involvement, cranial nerve involvement, and parenchymal involvement (most common). Pachymeningeal disease often takes the form of pachymeningeal thickening with homogeneous enhancement. Leptomeningeal involvement shows leptomeningeal enhancement, particularly around basal aspects of brain and circle of Willis. The manifestation of parenchymal involvement includes periventricular high-T2 signal white matter lesions and enhancing masses or nodule.

**FIGURE 6.7**   Neuro-behçet disease in a 35-year-old woman with head pain, somnolence, glossolalia, and oral ulcer. (A, B) Brain MRI demonstrates T2-FLAIR hyperintensity within bilateral posterior limb of internal capsule and right cerebral peduncles (*arrows*).

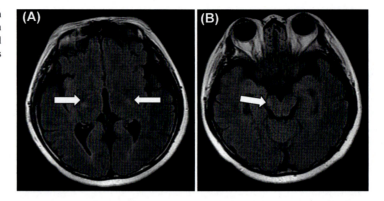

**FIGURE 6.8**   Behçet disease in a 24-year-old man with right lower quadrant pain and oral ulcer. (A) Noncontrast abdominal CT scan showed marked wall thickening (arrow) in cecum and terminal ileum. (B) Photograph obtained at colonoscopy shows diffusely scattered punched-out ulcers (*arrows*) and multiple polypi (*arrowheads*) in cecum and terminal ileum.

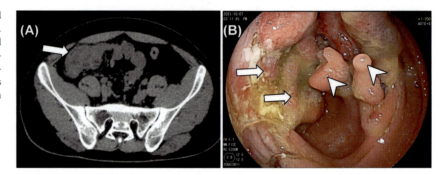

## 18. Behçet disease

Behçet disease is a multisystemic and chronic inflammatory vasculitis of unknown etiology. The classic clinical triad of Behçet disease consists of oral ulceration, genital ulceration, and ocular manifestations. Additional clinical manifestations include skin, joints, GI tract, CNS, cardiovascular system, etc. The underlying pathologic process in Behçet disease is vasculitis and perivascular inflammatory infiltrates affecting vessels of differing sizes in various organs. In 3% of cases, CNS manifestations occur first, making diagnosis significantly more challenging. Signs and symptoms include headaches, sensory disturbances, personality changes, dysarthria, and cerebellar signs [4]. Neuro-Behçet disease (NBD) has a wide variety of manifestations in the CNS, including focal or multifocal lesions, meningoencephalitis, and cerebral vein thrombosis. Lesions in NBD typically involve the brainstem (most common, and typically in pons), basal ganglia (bilateral in one-third of cases), and thalamus [55] (Fig. 6.7).

The well-known MRI findings in NBD are small foci of high signal intensity on T2-weighted images; these foci are iso- or hypointense relative to brain parenchyma on T1-weighted images. Lesions may be circular, linear, crescent shaped, or irregular. The development and disappearance of lesions at CT and MRI correlate with the course of clinical neurologic deficits [56]. Parenchymal Behçet disease seen as a space-occupying lesion or masquerading as unilateral brain tumor is an atypical manifestation. Behçet disease affects GI tract in 5%–60% of patients; this results in lymphocytic vasculitis in venules of bowel wall, with consequent chronic inflammation and intestinal ulceration [57]. The presence of ulcers is the radiologic and pathologic hallmark of Behçet disease, and ulcers are most commonly located in the ileocecal region (Fig. 6.8). In addition to ulcers, Behçet disease may cause esophageal ulceration, intestinal lymphangiectasia, and ulceration of stomach, duodenum, colon, rectum, or anus, with perianal fistulas, pancreatitis, and celiac sprue [58,59].

## 19. Systemic sclerosis

Systemic sclerosis, or scleroderma, is an autoimmune connective tissue disorder characterized by multisystem fibrosis and soft-tissue calcification. The disease is characterized by the widespread deposition of collagen and other extracellular matrix proteins. This is believed to occur as a result of abnormal immune response. Small vessels are involved early in the disease, accounting for involvement of organs with a dense capillary network. This results in perivascular fibrosis and gradual luminal stenosis [60]. Nearly 90% of patients with systemic sclerosis have evidence of GI involvement, which is, ultimately, a substantial cause of morbidity [61]. Esophageal involvement typically affects the distal two-thirds of the esophagus due to lack of striated muscle in upper one-third. Findings of esophageal involvement include decreased or absent esophageal peristalsis combined with prominent gastroesophageal reflux from incompetent lower esophageal sphincter. Esophagitis is frequently present, and associated complications such as esophageal stricture or Barrett metaplasia are fairly common. Gastric involvement is relatively uncommon but can result in delayed gastric emptying with or without gastric dilatation. Gastric vascular antral ectasia (dilated submucosal capillaries), often known as watermelon stomach, may also occur. The small bowel is affected in more than 60% of scleroderma patients, duodenum most frequently. Small bowel findings include hypomotility from smooth muscle atrophy and fibrosis, which leads to stasis, dilatation, and pseudoobstruction. "Hide-bound" sign of valvular packing is a fairly specific finding and may be seen in as many as 60% of patients with scleroderma. Hepatobiliary manifestations of scleroderma are only present symptomatically in a minority of patients. Around 2.5% of patients with scleroderma develop clinically significant primary biliary cholangitis. However, antibody test suggest subclinical disease may be present in as many as 15% of patients [62]. The severity of CNS involvement in patients with diffuse cutaneous systemic sclerosis ranges from minor (e.g., headache and radiculopathy) to life threatening (e.g., stroke, seizures, vasculitis, and myelopathy). CNS vasculitis is a rare complication of diffuse cutaneous systemic sclerosis that is most often diagnosed without brain biopsy and solely on the basis of imaging findings [63]. Improvement in patients with CNS involvement has been observed with administration of immunosuppressive therapy.

## 20. Amyloidosis

Amyloidosis is a systemic disease that is characterized by deposition of fibrillary proteins in different organs. The disease may have a male predilection. It typically affects middle-aged individuals around 60 years. Symptoms depend on the site of protein deposition. Most common sites are the kidneys, heart, liver, and GI tract. GI amyloidosis is relatively common, although symptomatic involvement is more rare. It is diagnosed if there are persistent GI symptoms with endoscopic biopsy-proven amyloid deposition. Weight loss (most common) and GI bleeding are main symptoms. Other symptoms include gastro-esophageal reflux, constipation, nausea, diarrhea, weight loss, early satiety, and abdominal pain. The most frequently affected site in GI amyloidosis is the duodenum [64], followed by the stomach, colon and rectum, and esophagus.

Radiological findings are rare and nonspecific, unlike pathologic high specificity. On fluoroscopy, gastric mucosa may display thickened folds, which may appear nodular or mass-like and which may contain calcifications. CT features include diffuse wall thickening of involved segment of bowel, intussusception, dilatation, or luminal narrowing. The dilatation depends upon the degree of hypomotility, whereas luminal narrowing resulted from either amyloid infiltration or secondary to ischemia. Liver involvement in amyloidosis is uncommon. There is amyloid deposition in liver parenchyma which occurs along sinusoids within the space of Disse or in blood vessel walls. Hepatocytes can be markedly compressed by extensive accumulation of amyloid and they may atrophy or nearly disappear. Hepatomegaly and heterogeneous appearances of liver parenchyma are the most common, though nonspecific, imaging features. CT patterns of hepatic amyloidosis include hepatomegaly, parenchymal masses, and calcifications that may be focal, lobar, or diffuse. The diffuse hypo-attenuation of liver parenchyma mimics hepatic steatosis. Parenchymal heterogeneity and a mottled pattern of contrast enhancement are likely the result of impaired blood flow due to vascular amyloid deposition [65,66].

Cerebral amyloid deposition diseases are a group of related conditions characterized by the accumulation of cerebral amyloid-β in various parts of CNS. They include Alzheimer disease (AD), cerebral amyloid angiopathy (CAA), CAA-related inflammation, and cerebral amyloidoma. AD is the most common cerebral amyloid deposition disease. CT and MRI are predominantly performed to exclude alternative causes of mild cognitive impairment or dementia, particularly in early stages of AD, and to quantify volume loss. As AD progresses, structural changes may become apparent at CT and MRI, including atrophy of entorhinal cortex, hippocampus, and precuneus. CAA is an important cause of primary intracerebral hemorrhage (ICH) in older patients. It may manifest sporadically, in association with AD, or rarely in its hereditary form. The disease is characterized by deposition of β-amyloid in media and adventitia of small arteries within brain and leptomeninges. Features of CAA at CT and MRI as defined by modified Boston Criteria include multiple ICHs or cerebral microbleeds restricted to lobar, cortical, or cortical—subcortical regions, or alternatively a single hemorrhage or microbleed in these regions in the presence of cortical superficial siderosis [67]. Susceptibility-weighted imaging is valuable for the detection of cerebral microbleeds, which appear as round hypointense dots. The location of hemorrhage is a key differentiator of CAA-associated ICH from more common ICH caused by hypertension, which classically appears as deep hemorrhage centered in putamen (basal ganglia), thalami, pons, and cerebellum. Inflammatory CAA comprises two histopathologically distinct entities characterized by inflammatory response to β-amyloid cerebrovascular deposition. In addition to multiple cortical microbleeds seen with CAA, MRI may also show infiltrative or tumefactive white matter lesions and patchy leptomeningeal or parenchymal enhancement, which help to distinguish this disease process from CAA without inflammation [68]. Cerebral amyloidoma is a rare form of focal β-amyloid deposition that manifests in middle age with symptoms of headache, seizures, and focal deficits. The disease is characterized by solitary or multiple supratentorial solid masses centered in white matter, often extending medially to ependyma of lateral ventricle, with surrounding edema and irregular enhancing margins [69]. At CT, the mass is typically showing iso- to hyper-attenuating and intense contrast enhancement.

# References

[1] Bleibel W, Al-Osaimi AM. Hepatic encephalopathy. Saudi J Gastroenterol 2012;18(5):301–9.

[2] Córdoba J. New assessment of hepatic encephalopathy. J Hepatol 2011;54(5):1030–40.

[3] Rovira A, Alonso J, Córdoba J. MR imaging findings in hepatic encephalopathy. AJNR Am J Neuroradiol 2008;29(9):1612–21.

[4] Hegde AN, Mohan S, Lath N, et al. Differential diagnosis for bilateral abnormalities of the basal ganglia and thalamus. Radiographics 2011;31(1):5–30.

[5] Stracciari A, Guarino M, Pazzaglia P, et al. Acquired hepatocerebral degeneration: full recovery after liver transplantation. J Neurol Neurosurg Psychiatr 2001;70(1):136–7.

[6] Lee J, Lacomis D, Comu S, et al. Acquired hepatocerebral degeneration: MR and pathologic findings. AJNR Am J Neuroradiol 1998;19(3):485–7.

[7] Huang FZ, Hou X, Zhou TQ, et al. Hepatic encephalopathy coexists with acquired chronic hepatocerebral degeneration. Neurosciences 2015;20(3):277–9.

[8] Weathers AL, Lewis SL. Rare and unusual... or are they? Less commonly diagnosed encephalopathies associated with systemic disease. Semin Neurol 2009;29:136–53.

[9] Zhang XP, Tian H. Pathogenesis of pancreatic encephalopathy in severe acute pancreatitis. Hepatobiliary Pancreat Dis Int 2007;6:134–40.

[10] Ding X, Liu CA, Gong JP, Li SW. Pancreatic encephalopathy in 24 patients with severe acute pancreatitis. Hepatobiliary Pancreat Dis Int 2004;3:608–11.

[11] Sharma V, Sharma R, Rana SS, Bhasin DK. Pancreatic encephalopathy: an unusual cause of asterixis. JOP 2014;15:383–4.

[12] Sharf B, Bental E. Pancreatic encephalopathy. J Neurol Neurosurg Psychiatry 1971;34:357–61.

[13] Lewis M, Howdle PD. The neurology of liver failure. QJM 2003;96(9):623–33.

[14] Maeda H, Sato M, Yoshikawa A, et al. Brain MR imaging in patients with hepatic cirrhosisrelationship between high intensity signal in basal ganglia on T1-weighted image s and elemental concentrations in brain. Neuroradiology 1997;39(8):546–50.

[15] Weissenborn K, Bokemeyer M, Krause J, et al. Neurological and neuropsychiatric syndromes associated with liver disease. AIDS 2005;19(Suppl. 3):S93–8.

[16] Wang MQ, Dake MD, Cui ZP, et al. Portal-systemic myelopathy after transjugular intrahepatic portosystemic shunt creationreport of four cases. J Vasc Intervent Radiol 2001;12(7):879–81.

[17] Upton MP, Pai RK, Vieth M, et al. Esophageal disease and pathology. Ann N Y Acad Sci 2011;1232:376–80.

[18] Mendoza G, Marti-Fabregas J, Kulisevsky J, et al. Hepatic myelopathya rare complication of portacaval shunt. Eur Neurol 1994;34(4):209–12.

[19] Lewis MB, MacQuillan G, Bamford JM, Howdle PD. Delayed myelopathic presentation of the acquired hepatocerebral degeneration syndrome. Neurology 2000;54(4):10–1.

[20] Pinarbasi B, Kaymakoglu S, Matur Z, et al. Are acquired hepatocerebral degeneration and hepatic myelopathy reversible? J Clin Gastroenterol 2009;43(2):176–81.

[21] Thomson AD, Marshall EJ. The natural history and pathophysiology of Wernicke's Encephalopathy and Korsakoff's Psychosis. Alcohol Alcohol 2006;41(2):151–8.

[22] Degnan AJ, Levy LM. Neuroimaging of rapidly progressive dementias, part 2: prion, inflammatory, neoplastic, and other etiologies. AJNR Am J Neuroradiol 2014;35(3):424–31.

[23] Diakou M, Kostadima V, Giannopoulos S, et al. Cerebral venous thrombosis in an adolescent with ulcerative colitis. Brain Dev 2011;33(1):49–51.

[24] Gobbelé R, Reith W, Block F. Cerebral vasculitis as a concomitant neurological illness in Crohn's disease. Nervenarzt 2000;71(4):299–304.

[25] Lam A, Borda IT, Inwood MJ, et al. Coagulation studies in ulcerative colitis and Crohn's disease. Gastroenterology 1975;68(2):245.

[26] Geissler A, Andus T, Roth M, Kullmann F, Caesar I, Held P, et al. Focal white-matter lesions in brain of patients with inflammatory bowel disease. Lancet April 8, 1995;345(8954):897–8.

[27] Lorincz MT. Neurologic Wilson's disease. Ann N Y Acad Sci 2010;1184:173–87.

[28] King AD, Walshe JM, Kendall BE, Chinn RJ, Paley MN, Wilkinson ID, et al. Cranial MR imaging in Wilson's disease. Am J Roentgenol. 1996;167(6):1579–84.

[29] Yu XE, Gao S, Yang RM, Han YZ. MR imaging of the brain in neurologic Wilson disease. Am J Neuroradiol. 2019;40(1):178–83.

[30] Kim TJ, Kim IO, Kim WS, Cheon JE, Moon SG, Kwon JW, et al. MR imaging of the brain in Wilson disease of childhood: findings before and after treatment with clinical correlation. Am J Neuroradiol 2006;27(6):1373–8.

[31] Jacobs DA, Markowitz CE, Liebeskind DS, et al. The "double panda sign" in Wilson's disease. Neurology 2003;61(7):969.

[32] Runge VM, Clanton JA, Smith FW, et al. Nuclear magnetic resonance of iron and copper disease states. Am J Roentgenol 1983;141(5):943–8.

[33] Umeoka S, Koyama T, Miki Y, et al. Pictorial review of tuberous sclerosis in various organs. Radiographics 2008;28(7):e32.

[34] Logue LG, Acker RE, Sienko AE. Best cases from the AFIP: angiomyolipomas in tuberous sclerosis. Radiographics 2003;23(1):241–6.

[35] Bernauer TA. The radial bands sign. Radiology 1999;212(3):761–2.

[36] Cha I, Cartwright D, Guis M, et al. Angiomyolipoma of the liver in fine-needle aspiration biopsies: its distinction from hepatocellular carcinoma. Cancer 1999;87(1):25–30.

[37] Damaskos C, Garmpis N, Garmpi A, et al. Angiomyolipoma of the liver: a rare benign tumor treated with a laparoscopic approach for the first time. Vivo 2017;31(6):1169–73.

[38] Vortmeyer AO, Lubensky IA, Skarulis M, et al. Multiple endocrine neoplasia type 1: atypical presentation, clinical course, and genetic analysis of multiple tumors. Mod Pathol 1999;12(9):919–24.

[39] Dreijerink KMA, Timmers HTM, Brown M. Twenty years of menin: emerging opportunities for restoration of transcriptional regulation in MEN1. Endocr Relat Cancer 2017;24(10):T135–45.

[40] Gianani R. The multiple endocrine neoplasia type-1 (MEN-1) syndrome and its effect on the pancreas. J Clin Endocrinol Metabol 2007;92(3):811–2.

[41] Leung RS, Biswas SV, Duncan M, et al. Imaging features of von Hippel-Lindau disease. Radiographics 2008;28(1):65–79.

[42] Hough DM, Stephens DH, Johnson CD, et al. Pancreatic lesions in von Hippel-Lindau disease: prevalence, clinical significance, and CT findings. Am J Roentgenol 1994;162(5):1091–4.

[43] Maher E, Neumann H, Richard S. von Hippel–Lindau disease: a clinical and scientific review. Eur J Hum Genet 2011;19:617–23.

[44] Choyke PL, Glenn GM, Walther MM, et al. Von Hippel Lindau disease: genetic, clinical, and imaging features. Radiology 1995;194(3):629–42.

[45] Filling-Katz MR, Choyke PL, Oldfield E, et al. Central nervous system involvement in von Hippel-Lindau disease. Neurology 1991;41(1):41–6.

[46] Ianora AA, Memeo M, Sabba C, et al. Hereditary hemorrhagic telangiectasia: multi-detector row helical CT assessment of hepatic involvement. Radiology 2004;230(1):250–9.

[47] Memeo M, Stabile Ianora AA, Scardapane A, Buonamico P, Sabba C, Angelelli G. Hepatic involvement in hereditary hemorrhagic telangiectasia: CT findings. Abdom Imag 2004;29:211–20.

[48] Aoki S, Barkovich AJ, Nishimura K, et al. Neurofibromatosis types 1 and 2: cranial MR findings. Radiology 1989;172(2):527–34.

[49] Traill Z, Tuson J, Woodham C. Adrenal carcinoma in a patient with Gardner's syndrome: imaging findings. AJR Am J Roentgenol 1995;165(6):1460–1.

[50] Deshpande V, Zen Y, Chan JK, et al. Consensus statement on the pathology of IgG4-related disease. Mod Pathol Off J USA Canadian Acad Pathol Inc 2012;25(9):1181.

[51] Hirano K, Tada M, Isayama H, et al. Endoscopic evaluation of factors contributing to intrapancreatic biliary stricture in autoimmune pancreatitis. Gastrointest Endosc 2010;71(1):85–90.

[52] Fujita A, Sakai O, Chapman MN, Sugimoto H. IgG4-related disease of the head and neck: CT and MR imaging manifestations. Radiographics 2012;32(7):1945–58.

[53] Warshauer DM, Lee JK. Imaging manifestations of abdominal sarcoidosis. Am J Roentgenol 2004;182(1):15–28.

[54] Jung G, Brill N, Poll LW, et al. MRI of hepatic sarcoidosis: large confluent lesions mimicking malignancy. AJR Am J Roentgenol 2004;183(1):171–3.

[55] Matsuo K, Yamada K, Nakajima K, et al. Neuro-Behçet disease mimicking brain tumor. Am J Neuroradiol 2005;26(3):650–3.

[56] Patel DV, Neuman MJ, Hier DB. Reversibility of CT and MR findings in neuro-Behçet disease. J Comput Assist Tomogr 1989;13:669–73.

[57] ChungSY, Ha HK, Kim JH, et al. Radiologic findings of Behçet syndrome involving the gastrointestinal tract. Radiographics 2001;21:911–24. discussion 924–916.

[58] Anti M, Marra G, Rapaccini GL, et al. Esophageal involvement in Behçet's syndrome. J Clin Gastroenterol 1986;8:514–9.

[59] Kasahara Y, Tanaka S, Nishino M, Umemura H, Shiraha S, Kuyama T. Intestinal involvement in Behçet's disease: review of 136 surgical cases in the Japanese literature. Dis Colon Rectum 1981;24:103–6.

[60] Rohrmann Jr CA, Ricci MT, Krishnamurthy S, Schuffler MD. Radiologic and histologic differentiation of neuromuscular disorders of the gastrointestinal tract: visceral myopathies, visceral neuropathies, and progressive systemic sclerosis. Am J Roentgenol 1984;143(5):933–41.

[61] Akesson A, Wollheim FA. Organ manifestations in 100 patients with progressive systemic sclerosis: a comparison between the CREST syndrome and diffuse scleroderma. Br J Rheumatol 1989;28(4):281–6.

[62] Norman GL, Bialek A, Encabo S, et al. Is prevalence of PBC underestimated in patients with systemic sclerosis? Dig Liver Dis 2009;41(10):762–4.

[63] Ishida K, Kamata T, Tsukagoshi H, Tanizaki Y. Progressive systemic sclerosis with CNS vasculitis and cyclosporin A therapy. J Neurol Neurosurg Psychiatry 1993;56(6):720.

[64] Cowan AJ, Skinner M, Seldin DC, et al. Amyloidosis of the gastrointestinal tract: a 13-year, single-center, referral experience. Haematologica 2013;98(1):141–6.

[65] Mainenti PP, D'Agostino L, Soscia E, Romano M, Salvatore M. Hepatic and splenic amyloidosis: dual-phase spiral CT findings. Abdom Imag 2003;28(5):688–90.

[66] Ye L, Shi H, Wu HM, Wang FY. Primarily isolated hepatic involvement of amyloidosis: a case report and overview. Medicine (Baltim) 2016;95(52):e5645.

[67] Linn J, Halpin A, Demaerel P, et al. Prevalence of superficial siderosis in patients with cerebral amyloid angiopathy. Neurology 2010;74(17):1346–50.

[68] Kotsenas AL, Morris JM, Wald JT, Parisi JE, Campeau NG. Tumefactive cerebral amyloid angiopathy mimicking CNS neoplasm. AJR Am J Roentgenol 2013;200(1):50–6.

[69] Gandhi D, Wee R, Goyal M. CT and MR imaging of intracerebral myloidoma: case report and review of the literature. Am J Neuroradiol 2003;24(3):519–22.

# Chapter 7

# Excretory system

Pinggui Lei and Bin Huang

*Department of Radiology, The Affiliated Hospital of Guizhou Medical University, Guiyang, Guizhou, China*

## 1. Von Hippel-Lindau syndrome

Von Hippel-Lindau (VHL) syndrome is an inherited autosomal dominant disorder that affects multiple organ systems. The disease is characterized by the growth of cysts and/or tumors. Tumors can be benign or malignant. The most characteristic type of tumor in VHL is hemangioblastoma, which is a benign tumor composed of newly formed blood vessels. Cysts are also a very common manifestation of VHL, occurring in the kidneys, pancreas, and reproductive tract. Renal cell carcinoma (RCC) and pancreatic neuroendocrine tumors are also seen in VHL. Tumor of the endolymphatic sac of the inner ear can be seen in patients with VHL. VHL is autosomal dominant with high penetrance, resulting in early onset and high frequency of clinical manifestations. It is the most common inherited kidney cancer syndrome. Males and females are equally affected. The average age of onset is 26 years, and the most common age is 18−30 years. VHL can be classified as follows: Type 1 (without pheochromocytoma) and Type 2 (with pheochromocytoma); Type 2 is further classified as: Type 2A: pheochromocytoma is present along with CNS hemangioblastomas but no RCC; Type 2B: pheochromocytoma is present along with both CNS hemangioblastomas and RCC; Type 2C: pheochromocytoma is present without hemangioblastomas or RCC. Mutations in the VHL tumor suppressor gene located on chromosome 3 cause VHL. These mutations prevent VHL protein production or lead to abnormal production. Treatment of VHL syndrome depends on the location and size of the lesions and the extent of the disease. The manifestation of VHL was summarized in Table 7.1.

Multicentric renal cysts and clear cell RCCs may manifest in more than two-thirds of VHL patients. The cystic lesions may be simple benign cysts, complex atypical cysts with epithelial hyperplasia/cytologic atypia, or cystic RCC. The number and size of VHL cysts did not appear to correlate with malignant potential. Patients may be asymptomatic, although they present with massive cysts. On CT, solid RCCs shows heterogeneity with obvious early enhancement, followed by washout in the delayed phase. At MRI, solid RCCs show T1 hypointense, although the presence of hemorrhage may result in T1 hyperintense. In out-of-phase chemical shift MRI, the presence of intracellular lipid in clear cell RCCs may rarely result in signal loss. Tumors are typically T2 hyperintense and show obvious heterogeneous enhancement. Cystic RCCs may be manifested as cysts with enhancing solid components and/or thick nodular septa. In contrast, simple renal cysts are homogeneously T1 hypointense and T2 hyperintense and lack enhancement on postcontrast images (Figs 7.1 and 7.2).

## 2. Tuberous Sclerosis Complex

Tuberous sclerosis complex (TSC) is a multisystem neurocutaneous genetic disorder with an incidence of 1 per 6000 to 10,000 live births. TSC is known to have a variable presentation that classically involves the brain, skin, kidneys, heart, eyes, and lungs, but can affect any organ system. The hallmark of the disease is tumors consisting of glial−neuronal and retinal hamartomas, subependymal giant cell tumors, cardiac rhabdomyomas, renal and extrarenal angiomyolipomas (AMLs), and pulmonary lymphangioleiomyomatosis. Despite the typically benign pathology of these tumors, they may lead to secondary outflow abnormalities due to mass effect (i.e., hydrocephalus, renal, and cardiac dysfunction) or interruption of the function of the normal tissue (i.e., seizures and arrhythmia). The clinical manifestations of TSC are the result of dysfunction in cell differentiation, proliferation, and migration during early fetal development. TSC is known to be caused by a mutation in either the *TSC1* or *TSC2* gene. The most common signs and symptoms of TSC are the triad

*Multi-system Imaging Spectrum associated with Neurologic Diseases.* https://doi.org/10.1016/B978-0-323-91795-7.00001-4

**TABLE 7.1** VHL manifestation.

| Retina | |
|---|---|
| | Hemangioblastomas (HBs) |
| CNS | |
| | Cerebellar and spinal HBs |
| Head and neck | |
| | Endolymphatic sac tumors |
| Pancreas | |
| | Pancreatic cysts |
| | Serous cystadenomas |
| | Pancreatic Neuroendocrine tumors (NETs) |
| Kidney | |
| | Renal cysts |
| | Clear cell Renal cell carcinomas |
| Adrenal gland | |
| | Pheochromocytoma |
| Reproductive organs | |
| | Epididymal cysts |
| | Papillary cystadenoma of epididymis |
| | Broad ligament cystadenoma |

(facial angiofibroma, seizures, and mental retardation). At present, 75% of all patients have facial angiofibroma, 90% have seizures, and about 50% have mental retardation (Fig. 7.3).

Renal AMLs are the most common benign tumors of the kidney, characterized by variable amounts of abnormal vessels, immature smooth-muscle and fat cells. CT permits specific diagnosis of a renal AML by demonstrating the presence of intratumoral fat. Typical CT findings of renal AMLs are noncalcified cortical tumors containing fat of less than −20 HU. Although unenhanced CT with thin sections is useful for detecting small amounts of fat, intratumoral fat cannot be detected in approximately 4.5% of all renal AMLs. These tumors frequently become clinically problematic because differentiating them from RCCs is difficult. Jinzaki et al. suggested that AMLs with minimal fat should be considered when the tumor demonstrates a) hyperattenuation at unenhanced CT, b) homogeneous enhancement, c) hypointensity on T2WI, and d) homogeneous isoechogenicity at ultrasonography.

Renal cysts or polycystic kidney disease can develop in patients with TSC. As opposed to renal AMLs, renal cysts occur in younger children. Although renal cysts are generally asymptomatic, they can more frequently cause subsequent hypertension or renal failure than can renal AMLs. CT findings in RCCs depend on their subtypes, owing to microvessel density or the presence of intratumoral necrosis or hemorrhage. Clear cell carcinoma, which is the most common subtype of RCC, typically demonstrates heterogeneous enhancement and early washout at biphasic contrast-enhanced CT. Chromophobe RCCs frequently demonstrate early weak enhancement and early washout. Papillary carcinomas tend to exhibit gradual enhancement, which is sometimes difficult to differentiate from that of AMLs with minimal fat (Fig. 7.4).

## 3. Behçet's disease

Behçet's disease is a kind of disease based on vascular inflammation pathological changes of chronic, relapsing auto-immune/inflammatory bowel disease, mainly characterized by recurrent oral ulcers, genital ulcers, uveitis, and skin lesions; it also affects the surrounding blood vessels, heart, nervous system, gastrointestinal tract, joints, lungs, kidneys, and other

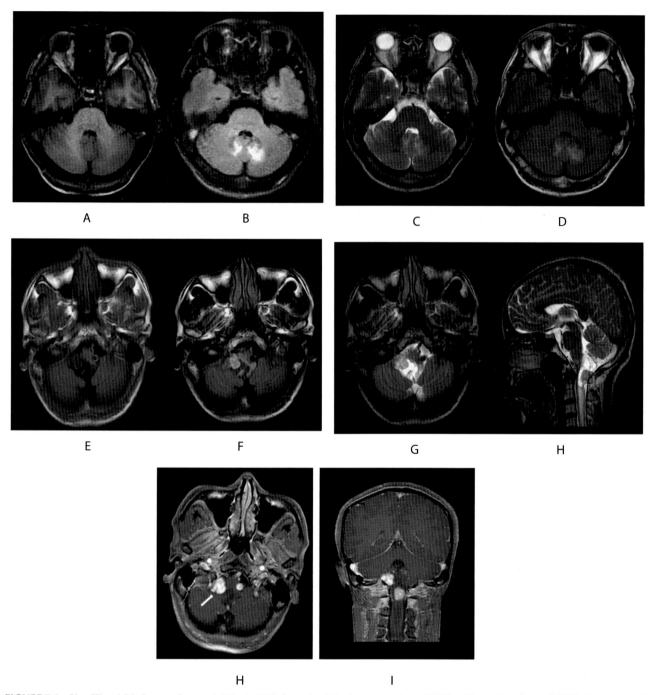

**FIGURE 7.1  Von Hippel-Lindau syndrome.** A-H brain MRI showed nodular hypo-intensity on $T_1$WI and hyper-intensity on $T_2$WI in the vermis and medulla oblongata of cerebellum, H−I contrast-enhanced MRI images show an avidly enhancing nodule which closes to the cerebellopontine angle.

organs. The onset age is mostly 15−50 years; median age of onset is 34 years. The incidence of male and female patients is similar, but early onset of male patients is more likely to be involved in important organs, and prognosis is poor. At present, the cause of the disease is not completely clear and may be related to heredity, such as the human leukocyte antigen (HLA) B51 gene, infection (some patients may be associated with tuberculosis), and living environment. High

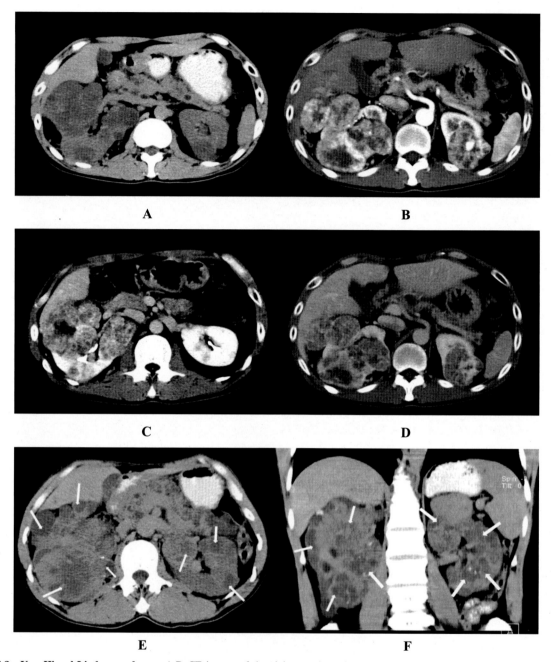

**FIGURE 7.2  Von Hippel-Lindau syndrome**. A-D CT images of the abdomen showed abnormal density shadow in kidneys. E-F after contrast enhancement, the lesion showed no obvious enhancement.

prevalence, familial aggregation, and sibling recurrence rates in specific geographic regions and ethnic populations are closely associated with HLA-B51, suggesting genetic factors play a key role in susceptibility.

## 4. Uremic encephalopathy

Uremia patients have abnormal manifestations of the central nervous system such as nerves and spirits, which are called uremic encephalopathy, also called renal encephalopathy. It belongs to the category of metabolic encephalopathy, which

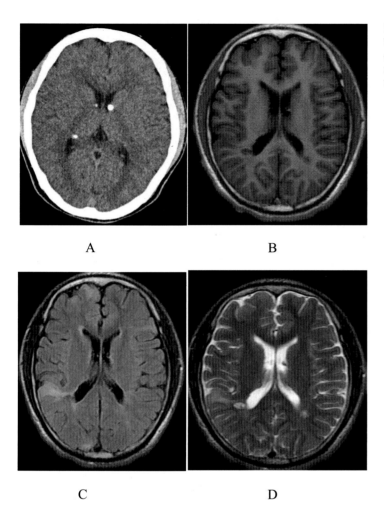

A

B

C

D

**FIGURE 7.3 Tuberous sclerosis complex** A. CT scan of the brain showed multiple nodular high-density shadows under the ependyma on both sides. B-D. Brain MRI showed multiple nodular hypo-intensity on $T_1WI$ and hyper-intensity on $T_2WI$ under the ependyma on both sides.

mainly manifests as acute or subacute reversible neurological and psychiatric symptoms, which may be caused by uremic toxin accumulation, nutritional deficiency, metabolic disorders, dialysis hypertension, imbalance syndrome, transplant rejection, and drugs (antibiotics and antiepilepsy drug use), including some stress states, such as trauma, infection, etc. Uremic toxins such as small molecular toxins (chlorine, guanidine, amines, phenols, etc.), medium molecular substances, and macromolecular parathyroid hormone accumulate in the blood and brain, inhibiting normal metabolism of brain cells. The clinical manifestations of uremic encephalopathy (UE) are complex and diverse, including headache, dyskinesia, delirium, coma and even epileptiform seizures in severe cases, and lack of specificity, which bring certain difficulties to diagnosis. Most UEs can be reversed with brain damage after timely treatment. Therefore, early diagnosis is of great significance to the prognosis of patients.

## 5. Polycystic kidneys disease

Polycystic kidney disease is the main cause of end-stage renal failure and a common indication of dialysis or kidney transplantation. Polycystic nephropathy may occur occasionally as a developmental abnormality or may occur in adulthood, but most forms are hereditary. In acquired forms, simple cysts can form in the kidneys due to

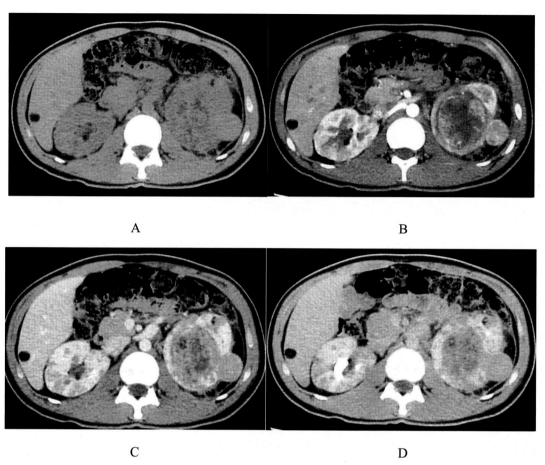

A

B

C

D

**FIGURE 7.4** **Tuberous sclerosis complex** A. CT scan of the abdomen showed abnormal density shadow in liver and renal. B-D. After contrast enhancement, the lesion showed obvious enhanced with inhomogeneity.

age; dialysis, drugs, and hormones can lead to polycystic diseases; and renal cysts are usually a secondary manifestation of hereditary proliferation syndrome. Hereditary polycystic nephropathy is caused by mutations in a single gene lineage, inherited as Mendel traits, including autosomal dominant and autosomal recessive polycystic nephropathy, renal tuberculosis, and myelin cystic disease. The age of onset, the severity of symptoms, and the rate of progression to end-stage renal failure or death vary widely. Autosomal dominant hereditary polycystic nephropathy is the most common polycystic kidney disease. Family-induced systemic artery disease caused by polycystic kidneys is a hereditary kidney disease that affects intracranial aneurysms, intracranial artery peeling, and autosomal dominant polycystic nephropathy in one in every 400 to 1000 people. In patients with autosomal dominant genetic polycystic nephropathy, the use of magnetic resonance angiography to screen for intracranial aneurysms is a cost-effective method. Polycystic kidney progression to renal failure can cause encephalopathy or coma accompanied by excessive ventilation and obvious involuntary movements, including tremors, flapping-like tremors, myoclonics, and hand-foot convulsions. Seizures or total seizures and symptoms of lesions are common and can occur in the epithelial cortex or in the debrain straight posture. Medication is not effective. Patients with long-term dialysis may develop bilateral basal section lesions syndrome (Fig. 7.5).

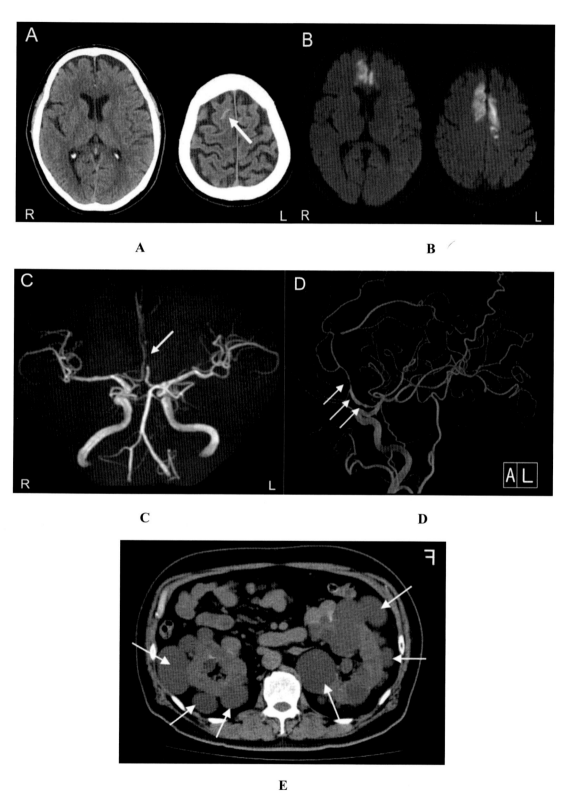

**FIGURE 7.5   A-E kidney encephalopathy caused by polycystic kidneys.** Polycystic kidney progression to renal failure can cause encephalopathy or coma accompanied by excessive ventilation and obvious involuntary movements, including tremors, flapping-like tremors, myoclonics, and hand-foot convulsions. Seizures or total seizures and symptoms of lesions are common and can occur in the epithelial cortex or in the debrain straight posture. Medication is not effective. Patients with long-term dialysis may develop bilateral basal section lesions syndrome.

## Further reading

[1] Ganeshan D, Menias CO, Pickhardt PJ, et al. Tumors in von Hippel-lindau syndrome: from head to toe-comprehensive state-of-the-art review. Radiographics 2018;38(3):849–66.

[2] Kanno H, Kobayashi N, Nakanowatari S. Pathological and clinical features and management of central nervous system hemangioblastomas in von Hippel-lindau disease. J Kidney Cancer VHL 2014;1(4):46–55.

[3] VHL Family Alliance. The VHL handbook: what you need to know about VHL—a reference handbook for people with von Hippel-Lindau disease, their families, and support personnel. Boston, Mass: CreateSpace (Amazon's independent publishing platform); 2014.

[4] Keutgen XM, Hammel P, Choyke PL, Libutti SK, Jonasch E, Kebebew E. Evaluation and management of pancreatic lesions in patients with von Hippel-Lindau disease. Nat Rev Clin Oncol 2016;13(9):537–49.

[5] Yang B, Autorino R, Remer EM, et al. Probe ablation as salvage therapy for renal tumors in von Hippel-Lindau patients: the Cleveland Clinic experience with 3 years follow-up. Urol Oncol 2013;31(5):686–92.

[6] Randle SC. Tuberous sclerosis complex: a review. Pediatr Ann 2017;46(4):e166–71.

[7] von Ranke FM, Faria IM, Zanetti G, Hochhegger B, Souza Jr AS, Marchiori E. Imaging of tuberous sclerosis complex: a pictorial review. Radiol Bras 2017;50(1):48–54.

[8] Kingswood JC, Bissler JJ, Budde K, et al. Review of the tuberous sclerosis renal guidelines from the 2012 consensus conference: current data and future study. Nephron 2016;134(2):51–8.

[9] Chae EJ, Do KH, Seo JB, et al. Radiologic and clinical findings of Behçet disease: comprehensive review of multisystemic involvement. Radiographics 2008;28(5):e31.

[10] Borhani-Haghighi A, Kardeh B, Banerjee S, et al. Neuro-Behcet's disease: an update on diagnosis, differential diagnoses, and treatment. Mult Scler Relat Disord 2019;39:101906.

[11] Topcuoglu OM, Topcuoglu ED, Altay CM, Genc S. Imaging pearls of pediatric Behçet's disease. Eur J Radiol 2017;94:115–24.

[12] Iwafuchi Y, Okamoto K, Oyama Y, et al. Posterior reversible encephalopathy syndrome in a patient with severe uremia without hypertension. Intern Med 2016;55(1):63–8.

[13] Canney M, Kelly D, Clarkson M. Posterior reversible encephalopathy syndrome in end-stage kidney disease: not strictly posterior or reversible. Am J Nephrol 2015;41(3):177–82.

[14] Kim DM, Lee IH, Song CJ. Uremic encephalopathy: MR imaging findings and clinical correlation. AJNR Am J Neuroradiol 2016;37(9):1604–9.

[15] Yoon JE, Kim JS, Park JH, et al. Uremic parkinsonism with atypical phenotypes and radiologic features. Metab Brain Dis 2016;31(2):481–4.

[16] Ahuja CK, Yadav MK, Khandelwal N. Mystery Case: syndrome of bilateral basal ganglia lesions in uremic encephalopathy. Neurology 2016;86(17):e182–3.

[17] Malhotra A, Wu X, Matouk CC, Forman HP, Gandhi D, Sanelli P. MR angiography screening and surveillance for intracranial aneurysms in autosomal dominant polycystic kidney disease: a cost-effectiveness analysis. Radiology May 2019;291(2):400–8.

[18] Tanaka M, Takasugi J, Hatate J, Otsuka N, Sugiura S, Itoh T, et al. Anterior cerebral artery dissection in a patient with autosomal dominant polycystic kidney disease. J Stroke Cerebrovasc Dis September 2019;28(9):e129–31.

# Chapter 8

# Reproductive system

**Xuntao Yin and Huiying Wu**
*Department of Radiology, Guangzhou Women and Children's Medical Center, Guangzhou, China*

## 1. Role of estrogen on neurological disorders

Three major naturally occurring forms of estrogens are estrone, estradiol (E2), and estriol, and E2 is the most potent and prevalent form. The actions of estrogens are mediated by estrogen receptors (ERs). The function of estrogen is to promote the development and maturity of sexual organs, to maintain secondary sexual characteristics, and to regulate endocrine function. It has been found that the function of estrogen is not limited to the reproductive system, it can also affect the nervous system, and has a certain neuroprotective effect.

### 1.1 Estrogen in the brain

Estrogen in the CNS is mainly from the peripheral system. As estrogen is fat soluble, it can easily be delivered to the CNS across the blood—brain barrier. In addition, a small amount might be synthesized from testosterone in the brain through the conversion of reactive astrocytes aromatization. Previous studies on the distribution of ER in CNS mainly focused on subcortical structure, especially hypothalamus. ER distributed in ventromedial nucleus of the hypothalamus may be involved in a variety of activities, including balance between endocrine and emotion, reproductive behavior, and cognitive function by affecting cerebral cortex. However, recent studies have found ER is widely distributed in the brain, including neocortices and hippocampus, as well as basal ganglia and amygdala underlying cortex [1]. The emerging notion that E2 and ER can act in multiple areas of brain led to an increased focus on its effects on neuronal physiology and neuroplasticity.

Estrogen and its receptors may be associated with neurological diseases. The emerging notion that E2 can act in multiple areas of the brain led to an increased focus on its effects on neuronal physiology and neuroplasticity. In vitro and in vivo studies indicated that E2 is a potent physiological modulator of CNS and participates in processes such as neurogenesis, regulation of neurotrophic factors expression, and regulation of antioxidant mechanisms. Women's premenstrual syndrome, postpartum depression, and postmenopausal depression have all been linked to low levels of estrogen in the body. Studies have shown estrogen replacement therapy can help prevent or treat these diseases [2].

There are three kinds of receptors for estrogen (Fig. 8.1). ERα and ERβ belong to ligand-activated nuclear transcription factors and are predominately present in nucleus and cytoplasm, with less than 2% on cellular membrane. Each ER exhibits differential tissue expression patterns, but both regulate gene transcription through classical genomic pathways, or by modulating cellular signaling pathways [3] such as the mitogen-activated protein kinases/extracellular signal-regulated kinases, modulation of intracellular calcium, cyclic adenosine monophosphate production, and regulation of phosphatidylinositol 3-kinase. G protein—coupled estrogen receptor 1 (GPER) was identified as the third ER. It is a membrane-associated ER associated with rapid estrogen-mediated effects. Over recent years, GPER emerged has a potential therapeutic target to induce neuroprotection, avoiding the side effects elicited by the activation of classical ERs [4]. The putative neuroprotection triggered by GPER selective activation was demonstrated in mood disorders, Alzheimer's disease (AD) or Parkinson's disease (PD) of male and female in vivo rodent models [3].

## 1.2 Estrogen and mood disorders

Mood disorders, including major depressive disorder (MDD), are a significant global health issue. Worldwide lifetime prevalence of mood disorders has been reported to be nearly 10% [5]. MDD is generally considered a brain-targeted disease associated with persistent sadness, guilt, anhedonia (reduced interest in rewarding stimuli), despair, and in some cases, suicide. Traditionally, MDD has been associated with deficiencies in serotonergic (5-HT), dopaminergic, and noradrenergic signaling within limbic, reward, and brainstem structures [6]. Recently, a substantial amount of research attention has been paid to the role of sex hormones, especially the steroid hormone estrogen, in driving development of MDD in women. Estrogens, especially E2, are known to induce dendritic spine plasticity and neuronal complexity, facilitate neurogenesis, regulate brain region volume and activity levels, and impact key neurotransmitter and growth factor systems implicated in depression. Several studies have noted increased depression among women taking hormonal contraceptives though collective findings generally suggest that contraception exerts minimal effects on mood [7]. The realization of neurobiological and behavioral effects of estrogen-containing treatments depends on numbers of factors including, but not limited to, age of the organism, etiology and duration of hormone depletion, type of estrogen, treatment route of administration, treatment regimen, and functional domain targeted [8]. Consideration of these factors is of key importance when assessing mood-impacting effects of this hormone. It has been suggested that genetic sex, estrogen, and the immune system significantly contribute to mood and mood disorders both individually and as converging, interactive factors [9]. Further to explore the complex interactions in the context of mental health is needed.

## 1.3 Estrogen and Alzheimer's disease

AD is the most common cause of dementia and the sixth leading cause of death in Western societies, affecting over 24 million patients worldwide [10]. AD affects more women than men, with a nearly 2:1 ratio in many countries, and with postmenopausal women accounting for over 60% of all those affected. Recent investigations targeting women of perimenopausal age demonstrated significant associations between menopause and biomarker indicators of increased AD risk. For instance, smaller medial temporal lobe volume was reported in surgical menopausal cases as compared to spontaneous menopause [11] and in recently postmenopausal women with subjective cognitive decline [12]. More direct evidence that the menopause transition is associated with AD risk comes from multimodality brain imaging studies reporting emergence of AD endophenotypes in midlife women carrying risk factors for AD, such as APOE-4 genotype and a family history of late-onset AD [13]. AD biomarker effects involved brain regions vulnerable to AD, such as posterior cingulate, precuneus, parieto-temporal, medial temporal, and frontal cortices (Fig. 8.2). These regions exhibit considerable overlap with the brain estrogen network (Fig. 8.3), further highlighting the connection between endocrine aging and cognitive aging in women.

## 1.4 Estrogen and Parkinson's disease

PD is a common neurodegenerative disease mainly characterized by α-synuclein pathology and loss of dopaminergic neurons in the substantia nigra pars compacta. The classic motor symptoms of PD are bradykinesia, rigidity, resting tremor, and postural and gait impairment. PD has also a constellation of nonmotor symptoms, including depression, anxiety, pain, orthostatic hypotension, and urinary, gastrointestinal, and sleep dysfunction, which can precede the motor features by more than a decade [14]. Epidemiologic studies support the idea of sex-related differences in risk factors for PD. Gonadal hormones and sex chromosomes might modulate disease risk by influencing epigenetic mechanisms. Preclinical evidence has suggested a potential neuroprotective effect of estrogens against dopaminergic damage through antiinflammatory, antioxidative, and antiapoptotic mechanisms, in addition to possible inhibitory effects on the formation and stabilization of α-synuclein fibrils—a key pathological feature of PD. Few studies have examined sex differences in imaging biomarkers for PD. In one MRI study that measured sex differences in brain structures in patients with PD [15], reduced cortical thickness in multiple brain regions including the frontal, parietal, temporal, and occipital lobes, associated with altered connectivity, was found in male patients in comparison with female patients. In a resting-state functional MRI study in drug-naive patients with early PD, sex-specific cortical and subcortical connectivity patterns within the sensorimotor network were reported, with connectivity being better preserved in women than in men [11], possibly related to sex-specific nigrostriatal dopaminergic pathways. PD symptoms seem to be influenced by the menstrual cycle. Worsening of PD symptoms can occur just before the onset of menses, when estrogen levels are reduced, whereas progressive improvement can be observed at the time of ovulation, when estrogen levels are higher. These findings support a positive effect of estrogens on the dopaminergic system [16]. One possible explanation might be the dopamine-sparing properties of estrogens, including inhibition of dopamine uptake, synthesis, and release [17].

## 1.5 Estrogen and multiple sclerosis

Multiple sclerosis (MS) is a chronic, progressive, demyelinating inflammatory autoimmune disease associated with a myriad of degenerative sensori/locomotor and cognitive deficits. Indeed, a woman's risk for developing MS increases after the pubertal transition, an effect linked to increasing levels of estrogens given that MS symptomology appears to decrease in intensity during the luteal phase of the menstrual cycle, when estrogen levels are low [18]. Additional clarity regarding the contributions of sex hormones alone and in combination is warranted as perplexingly; some studies note greater MS symptomology and worsened cognitive function in the premenstrual phase when sex hormones are generally at their lowest levels [19]. As well, though relapse rates increase significantly by 3 months postpartum, pregnancy is typically associated with symptom remission. Short-term corticosteroid treatment to manage MS symptoms during late pregnancy is considered safe with regards to fetal outcomes such as risk of preterm birth and low birth weight. Whether this treatment impacts affective outcomes in the pregnant or postpartum mother is not clear and represents an important area of investigation [9], given that corticosteroid treatments are known to induce psychiatric symptoms such as mania, depression, psychosis, and cognitive changes.

## 2. Paraneoplastic syndromes

Paraneoplastic neurological syndrome (PNS) is the most common reproductive system tumor in ovarian cancer and breast cancer [20]. The diagnosis of PNS in reproductive system tumors needs to exclude primary nervous system diseases, such as vascular and nervous diseases, infections, nervous system tumors, and genetic diseases. It is worth noting that typical PNS symptoms and specifically related antibodies are positive, and PNS can also be diagnosed without a primary tumor. The appearance of PNS symptoms can be earlier than the diagnosis of a primary tumor, and the interval is from years to weeks. PNS usually progresses subacutely within a few weeks, with different clinical manifestations. PNS associated with ovarian tumor can be manifested as peripheral polyneuropathies, such as generalized sensory loss or abnormality. The PNS of ovarian teratoma in young women or children is encephalitis, including mental disorder, memory disorder and cognitive behavior disorder, and even typical, motor disorder and autonomic nervous system disorder (Table 8.1). Some patients with ovarian cancer were positive for tumor autoantibodies but did not have any symptoms of PNS.

The most common antibody related to PNS is anticerebellar degeneration associated protein-2 (CDR2) antibody, also known as anti-Yo antibody. Positive anti-Yo antibody in serum indicates ovarian cancer, breast cancer, or other gynecological malignancies, and anti-RI indicates breast cancer; anti-N-Methyl-D-Aspartate receptor (NMDAR) antibody strongly suggests ovarian teratoma. The main purpose of imaging examination is to exclude other lesions that may cause neurological symptoms, such as primary brain tumors, metastases, and vascular diseases. In patients with PNS, MRI can indicate cerebellar atrophy and PET can indicate cerebellar hypometabolism. In 50% of patients with anti-NMDAR encephalitis, MRI can indicate high-density lesions in hippocampus, corpus callosum, temporal lobe, and frontal lobe (Figs. 8.1 and 8.2). Electroencephalogram is of great significance in the diagnosis of NMDAR encephalitis. Abnormal slow wave and pathological electrical activity can not only assist in the diagnosis of encephalitis, but also suggest adverse drug reaction to epilepsy.

In any case, the diagnosis of PNS requires further clinical search for primary tumors. When most patients have symptoms of right PNS, their primary tumor is in the early clinical stage, and the traditional diagnostic methods are easy to miss diagnosis. Pet should be more sensitive as a routine imaging method. PET/CT should be performed for those with relevant antibodies suggesting that the primary tumor is ovarian tumor. If no positive results are found in a series of primary screening, it should be rechecked every 6 months for at least 4 years to find the primary tumor.

**TABLE 8.1 PNS associated with gynecological cancers.**

Subacute cerebellar degeneration/cerebellar ataxia
Limbic encephalitis
Anti-NMDAR encephalitis
Opsoclonus-myoclonus syndrome
Subacute sensory neuropathy
Sensorimotor neuropathy
Lambert—eaton myasthenic syndrome
Myasthenia gravis

**FIGURE 8.1** MR images for anti-NMDAR encephalitis. Bilateral centrum semiovale subcortical and left cerebellar dentate nucleus showed T1WI slightly hyposignals, T2WI and T2WI-FLAIR slightly hypersignals, with no enhancement.

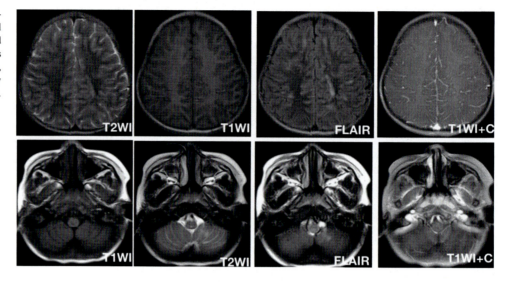

**FIGURE 8.2** MR images for anti-NMDAR encephalitis. Multiple patchy abnormal signals showed in bilateral basal ganglia, thalamus, cerebral peduncle, and right frontal lobe. T1WI showed slightly hyposignal, and T2WI and FLAIR showed slightly hypersignal, with no enhancement.

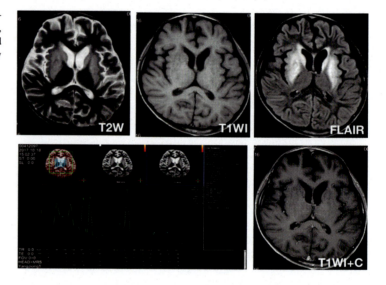

## 2.1 Subacute cerebellar degeneration/cerebellar ataxia

Paraneoplastic cerebellar degeneration (PCD) is the most frequent PNS in patients with breast or gynecologic tumors. PCD usually has an aggressive course and manifests as cerebellar dysfunction, provoking gait and truncal ataxia, dysarthria, and nystagmus. Diplopia and vertigo are also common. Any nonfamilial cerebellar ataxia that arises in patients older than 45 years should raise the suspicion of paraneoplastic syndrome.

The main neuropathologic finding in PCD is the loss of the Purkinje's cells with secondary astrogliosis and degeneration of the Purkinje's cell axons in the cerebellum and the cortex. This condition is caused by an autoimmune reaction. The antibody most commonly associated with PCD is anti-Yo, found with cancer of the ovary, breast, or other gynecological malignancies. Patients with PCD and positive anti-Yo antibodies almost always have underlying cancer; thus, an abdominal exploratory surgery should be considered for patients with cerebellar degeneration and positive anti-Yo with no evidence of cancer despite an extensive investigation because PCD can be induced even by a small tumor. Clinical signs usually develop in 2 months and then stabilize, leaving the patient very much disabled. PCD has been associated with ovarian carcinoma, fallopian tube carcinoma, and cervical cancer. Neurologic outcomes of patients with PCD are poor, and most patients remain disabled despite effective cancer treatment and immunotherapy.

## 2.2 Limbic encephalitis and anti-N-methyl-D-Aspartate receptor encephalitis

Limbic encephalitis consists of rapid development of neuropsychiatric symptoms including irritability, depression, seizures, hallucinations, and short-term memory loss and is characterized by degenerative. The exact cause of this syndrome is unknown. Various antibodies that react with neuronal cell intracellular antigens can cause limbic encephalitis. The antibodies more frequently associated with this disorder are anti-Hu, which can be found in ovarian tumors. Anti-NMDAR encephalitis is immune-mediated encephalitis usually affecting young females of reproductive age and associated with ovarian teratoma. This syndrome can be preceded by a viral prodrome of hyperthermia or headache. Anti-NMDAR encephalitis presents with neuropsychiatric and behavioral symptoms, accompanied by decreased level of consciousness, central hypoventilation, autonomic instability, and dyskinesias. It is potentially fatal. Most published cases demonstrated the presence of mature teratomas; however, immature teratomas account for about 20% of cases. Anti-NMDAR antibodies are produced by an immune response to neural tissues contained in teratomas. Diagnosis of anti-NMDAR encephalitis can be established by detection of these antibodies in the serum or cerebrospinal fluid (CSF), although they may sometimes only be detected in CSF. Intensive care and ventilator support may be required for a prolonged period time. Tumor resection is the cornerstone of treatment. Corticosteroids, plasmapheresis, immune-globulins, and anticonsultants can help alleviate the symptoms. Recovery may take months or years, and some patients experience permanent sequelae. However, as anti-NMDAR encephalitis results from reversible surface antibody-mediated neuronal dysfunction, most patients will improve or fully recover with treatment.

## 2.3 Opsoclonus-myoclonus syndrome

Opsoclonus is a subacute disorder of eye movements combined with ataxia. Patients with opsoclonus-myoclonus syndrome (OMS) present with opsoclonus, a disorder of eye movement characterized by involuntary conjugated saccades of large ample trunk and limbs. They can also develop cerebellar symptoms such as nystagmus, ataxia, and dysarthria. This syndrome can be idiopathic, metabolic, para-infectious, or paraneoplastic. The tumors more frequently involved in adults are small-cell lung cancer (SCLC), breast, and gynecological cancers. An investigation for cancer, with high-resolution CT of the chest and abdomen, mammography, and gynecological examination, should be performed. Currently recognized onco-neuronal antibodies are usually not found in paraneoplastic OMS, except anti-Ri when related to breast cancer. Treatment of OMS includes corticosteroids, immunotherapy, and treatment of the underlying cancer. Rarely is a good response obtained.

## 2.4 Paraneoplastic subacute sensory neuropathy

This paraneoplastic syndrome is most common associated with SCLC, but may be associated with other cancers such as cervical and ovarian carcinomas, and with uterine sarcoma. For example, in patients with stage I epithelial ovarian cancer, the incidence of subclinical sensory neuropathy is as high as 33.3%, and it becomes more clinically obvious in advanced-stage cancer where up to 71.1% of patients show neurophysiological abnormalities. Anti-Hu, amphipyisin, and anti-CV2 antibodies are sometimes implicated, but frequently no onconeural antibody is identified. Clinical manifestations initially include asymmetric numbness and pain of the limbs, and proprioception may also be affected. It can occur simultaneously with other paraneoplastic neurologic syndromes such as encephalomyelitis.

## 2.5 Sensorimotor neuropathy

The cornerstone of sensorimotor neuropathy is predominant motor impairment, usually of subacute onset. This paraneoplastic syndrome is frequently not associated with an identifiable onconeural antibody. Clinically apparent sensorimotor neuropathy is rarely associated with gynecologic tumors; however, patients with advanced stage epithelial ovarian cancer often present electrophysiological evidence of both sensory and motor involvement. It has also seldom been reported with squamous cell carcinoma of the cervix and uterine adenocarcinoma. Motor manifestations can be associated with CNS involvement, mostly encephalomyelitis, and autonomic dysfunction, leading to heterogeneous disorders. Electrophysiological studies show abnormalities in sensory conduction values.

## 2.6 Lambert–eaton myasthenic syndrome

The Lambert–Eaton myasthenic syndrome (LEMS) is defined as muscle weakness and fatigue that occur mostly in the pelvic girdle and thighs and is associated with dry mouth, dysarthria, dysphagia, blurred vision or diplopia, eyelid ptosis, paresthesia, and muscle pain. The syndrome has been associated with small-cell cervical cancer, ovarian cancer, and with uterine leiomyosarcoma. A diagnosis of LEMS mandates investigations to detect an underlying neoplasm. Clinically, patients present proximal muscle weakness, mostly affecting lower limbs, prominent fatigability, diminished deep tendon reflexes, ptosis, and anticholinergic symptoms such as dry mouth, dry eyes, postural hypotension, and constipation. LEMS is caused by antibodies directed against presynaptic P/Q type voltage-gated calcium channels at the neuromuscular junction, leading to impaired neuromuscular transmission. These antibodies are specific for LEMS and are detected in over 90% of cases, but do not predict the presence of cancer. Treatment cornerstone is directed against the underlying malignancy, as successful tumor treatment can be associated with improvement of LEMS. Options for symptomatic relief consist in pharmacologic modulation of acetylcholine and immunomodulation.

## 2.7 Myasthenia gravis

Myasthenia gravis (MG) is the most common neuromuscular transmission disorder. It is caused by antibodies directed against the postsynaptic nicotinic acetylcholine receptors, blocking the action of acetyl-choline and resulting in muscle weakness. Weakness in MG usually involves the face, neck, shoulders, and arms. Characteristically, extraocular muscles are affected, leading to diplopia and ptosis. Also, respiratory muscle weakness can precipitate acute primary respiratory failure. Elevated titer of serum binding antibodies to acetylcholine receptors confirms a clinical diagnosis of MG. Treatment of MG includes anticholinesterase medication, immunoglobulin therapy, and plasma exchanges. This condition is usually associated with thymic hyperplasia and thymoma but has also been rarely described with tumors of the female genital tract. Treatment of the underlying malignancy contributes to clinical improvement.

# 3. AIDS

## 3.1 AIDS patients with CNS cryptococcosis

*Cryptococcus neoformans* infection is a common opportunistic infection all over the world, and AIDS patients are at the highest clinical risk. Cryptococcal meningitis may be the first manifestation of HIV infection [21]. Patients who inhale spores or yeast cells from the respiratory tract are usually asymptomatic, and in most cases, the infection becomes localized in the lungs and causes pulmonary symptoms. It is reported that CD4 and CD8-mediated immunity is an important defense against cryptococcus invasion. In the cases with specific T cells damage, many virulence factors in cryptococcus neoformans, such as laccase, urokinase, phospholipase, melanin, can prompt it to live in the human body and replicate, which can lead to systemic spread disease, The CNS is a prone site of cryptococcus infection because there is no soluble cryptococcal antifactor in CSF.

AIDS-associated CNS cryptococcosis is usually subacute onset. The patients may have recurrent insidious headache, vomiting, fever, neck stiffness, changes in cognitive mental function, and even drowsiness. Severe headache may be attributed to two points: one is increased intracranial pressure; Secondly, cryptococcus proliferates in subarachnoid space, and various pathogenic factors gather to stimulate meningeal nerves to produce pain and cause meningeal irritation sign of varying degrees. Fever is also one of the common symptoms. Because AIDS patients do not have a strong inflammatory response, the body temperature only shows low or moderate fever [1]. Drowsiness and cognitive and mental function changes are mostly due to the invasion of the corresponding brain parenchyma, such as frontal lobe damage leading to memory and attention loss; temporal lobe damage leading mental disorders, such as delusions, hallucinations, delusions, and other symptoms; limbic lobe damage can lead to emotional disorders, abnormal behavior, and mental retardation.

The imaging manifestations of cryptococcosis in the CNS of AIDS patients are diversified, which is mainly closely related to the cryptococcal position of intracranial colonization and pathological changes. When cryptococcus spreads to the intracranial with blood and penetrated the walls of small blood vessels to colonize the perivascular space, namely Virchow—Robin space (VRS), no strong inflammatory response was produced, but only widening of VRS. The signal intensity of expanded VRS on all MR sequences was similar to that of CSF, with low signal intensity on T1WI, high signal intensity on T2WI, low signal intensity on FLAIR, low signal intensity on DWI, and no enhancement (Figs. 8.3 and 8.4). VRS widening is the earliest and most common manifestation of AIDS patients [22]. Therefore, CSF smear and culture should be performed as soon as possible for AIDS patients suspected to be complicated with cryptococcus infection if continuous level VRS expansion is present, so as to timely diagnose and prevent disease progression [23].

## 3.2 AIDS patients with CNS lymphoma

Primary CNS lymphoma (PCNSL) is a rare non-Hodgkin's lymphoma. The incidence of PCNSL is extremely low, accounting for only 2.2% of CNS tumors [24]. However, the incidence of PCNSL in AIDS patients is 3600 times higher than

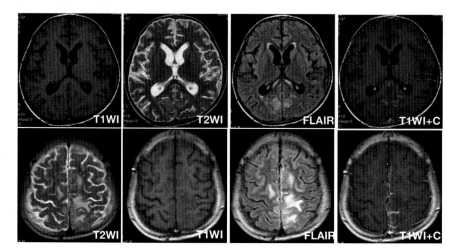

**FIGURE 8.3** Cryptococcosis in the CNS of AIDS. Patchy abnormal signal showed at the bilateral occipital/parietal/occipital lobes and splenium of corpus callosum. T1WI showed slightly hyposignal, T2WI and T2WI-FLAIR showed hypersignal with no enhancement.

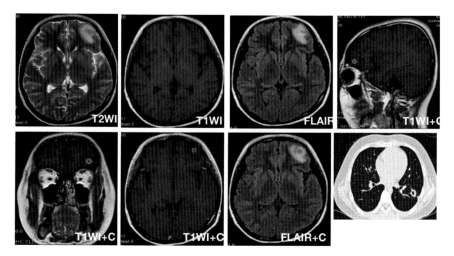

**FIGURE 8.4** Cryptococcosis in the CNS of AIDS. Left frontal lobe showed abnormal signal with hyposignal on T1WI, hypersignal on T2WI and FLAIR. T1WI-enhanced MRI showed ring enhancement. Chest CT showed pulmonary cryptococcal infection.

that in the general population, and about 2%–12% of AIDS patients can suffer from PCNSL, which is the most common intracranial mass lesion in AIDS patients except toxoplasma encephalitis. About 95% of PCNSL histological type is diffuse large B-cell lymphoma. Currently, AIDS-related PCNSL is believed to be associated with severe immune suppression and Epstein-Barr virus infection [25].

The clinical symptoms have no obvious specificity, and mainly include intracranial hypertension (headache, malignancy, visual impairment), neurocognitive dysfunction (lethargy, memory loss), focal neurological deficit (hemiplegia, speech impairment, cranial nerve palsy, and epilepsy) [26] and systemic symptoms. MRI manifestations of AIDS-related PCNSL are different from those of non-AIDS patients. Single supratentorial lesions are more common in non-AIDS patients, while cystic degeneration and necrosis are rarely seen. AIDS-related PCNSL shows supratentorial lesions and multiple lesions. It is usually found in the midline, mainly distributed in basal ganglia, corpus callosum, periventricular white matter area, frontal lobe and thalamus, sometimes confined to the subependymal area, or even completely located in the ventricle. The lesions can invade the ependymal, pia mater, or dura mater and spread along them.

# References

[1] Yague JG, Wang AC, Janssen WG, Hof PR, Garcia-Segura LM, Azcoitia I, et al. Aromatase distribution in the monkey temporal neocortex and hippocampus. Brain Res 2008;1209:115–27.

[2] Studd J. Hormone therapy for reproductive depression in women. Post Reprod Health 2014;20(4):132–7.

[3] Roque C, Mendes-Oliveira J, Duarte-Chendo C, Baltazar G. The role of G protein-coupled estrogen receptor 1 on neurological disorders. Front Neuroendocrinol 2019;55:100786.

[4] Prossnitz ER, Barton M. The G-protein-coupled estrogen receptor GPER in health and disease. Nat Rev Endocrinol 2011;7(12):715–26.

[5] Steel Z, Marnane C, Iranpour C, Chey T, Jackson JW, Patel V, et al. The global prevalence of common mental disorders: a systematic review and meta-analysis 1980−2013. Int J Epidemiol 2014;43(2):476−93.

[6] Krishnan V, Nestler EJ. The molecular neurobiology of depression. Nature 2008;455(7215):894−902.

[7] Robakis T, Williams KE, Nutkiewicz L, Rasgon NL. Hormonal contraceptives and mood: review of the literature and implications for future research. Curr Psychiatr Rep 2019;21(7):57.

[8] Engler-Chiurazzi EB, Brown CM, Povroznik JM, Simpkins JW. Estrogens as neuroprotectants: estrogenic actions in the context of cognitive aging and brain injury. Prog Neurobiol 2017;157:188−211.

[9] Engler-Chiurazzi EB, Chastain WH, Citron KK, Lambert LE, Kikkeri DN, Shrestha SS. Estrogen, the peripheral immune system and major depression—a reproductive lifespan perspective. Front Behav Neurosci 2022;16:850623.

[10] 2021 Alzheimer's disease facts and figures. Alzheimers Dement 2021;17(3):327−406.

[11] Zeydan B, Tosakulwong N, Schwarz CG, Senjem ML, Gunter JL, Reid RI, et al. Association of bilateral salpingo-oophorectomy before menopause onset with medial temporal lobe neurodegeneration. JAMA Neurol 2019;76(1):95−100.

[12] Conley AC, Albert KM, Boyd BD, Kim SG, Shokouhi S, McDonald BC, et al. Cognitive complaints are associated with smaller right medial temporal gray-matter volume in younger postmenopausal women. Menopause 2020;27(11):1220−7.

[13] Jett S, Malviya N, Schelbaum E, Jang G, Jahan E, Clancy K, et al. Endogenous and exogenous estrogen exposures: how women's reproductive health can drive brain aging and inform Alzheimer's prevention. Front Aging Neurosci 2022;14:831807.

[14] Postuma RB, Aarsland D, Barone P, Burn DJ, Hawkes CH, Oertel W, et al. Identifying prodromal Parkinson's disease: pre-motor disorders in Parkinson's disease. Mov Disord 2012;27(5):617−26.

[15] Yadav SK, Kathiresan N, Mohan S, Vasileiou G, Singh A, Kaura D, et al. Gender-based analysis of cortical thickness and structural connectivity in Parkinson's disease. J Neurol 2016;263(11):2308−18.

[16] Meoni S, Macerollo A, Moro E. Sex differences in movement disorders. Nat Rev Neurol 2020;16(2):84−96.

[17] Shulman LM. Gender differences in Parkinson's disease. Gend Med 2007;4(1):8−18.

[18] Moulton VR. Sex hormones in acquired immunity and autoimmune disease. Front Immunol 2018;9:2279.

[19] Guven Yorgun Y, Ozakbas S. Effect of hormonal changes on the neurological status in the menstrual cycle of patient with multiple sclerosis. Clin Neurol Neurosurg 2019;186:105499.

[20] Madhavan AA, Carr CM, Morris PP, Flanagan EP, Kotsenas AL, Hunt CH, et al. Imaging review of paraneoplastic neurologic syndromes. AJNR Am J Neuroradiol 2020;41(12):2176−87.

[21] Williamson PR, Jarvis JN, Panackal AA, Fisher MC, Molloy SF, Loyse A, et al. Cryptococcal meningitis: epidemiology, immunology, diagnosis and therapy. Nat Rev Neurol 2017;13(1):13−24.

[22] Kourbeti IS, Mylonakis E. Fungal central nervous system infections: prevalence and diagnosis. Expert Rev Anti Infect Ther 2014;12(2):265−73.

[23] Franco-Paredes C, Womack T, Bohlmeyer T, Sellers B, Hays A, Patel K, et al. Management of Cryptococcus gattii meningoencephalitis. Lancet Infect Dis 2015;15(3):348−55.

[24] Grommes C, DeAngelis LM. Primary CNS lymphoma. J Clin Oncol 2017;35(21):2410−8.

[25] Vangipuram R, Tyring SK. AIDS-associated malignancies. Cancer Treat Res 2019;177:1−21.

[26] Brandsma D, Bromberg JEC. Primary CNS lymphoma in HIV infection. Handb Clin Neurol 2018;152:177−86.

# Chapter 9

# Skeletal system

Daniel Phung, Gordon Crews, Raymond Huang and Nasim Sheikh-Bahaei
*Keck School of Medicine of USC, Los Angeles, CA, United States*

## 1. Ankylosing spondylitis

Ankylosing spondylitis (AS) is a seronegative inflammatory spondyloarthropathy afflicting approximately 0.1%−0.3% of the overall population [1,2]. The disease characteristically involves the outer capsule of the annulus fibrosus known as Sharpey's fibers attaching to adjacent vertebral body endplates. Inflammation of these fibers leads to erosive changes followed by ossification and multilevel fusion. Enthesitis also leads to enthesophyte formation. Sacroiliac joint involvement is common and other small joints may also be involved [3]. The disease has a predilection for patients who express human leukocyte antigen B27, an immune system presenting protein which binds to T-cells, and this population comprises 6% of all AS patients [1]. Patients often present in their teens to 20s with back and hip pain and stiffness, fatigue, and generalized joint pain. Fusion of the spine places patients at risk of serious neurologic injury even in the setting of minor trauma. Ground-level falls can cause unstable spine injury ranging from subtle fractures involving anterior and posterior columns extending through ossified disc space or widely displaced spinal fractures known as "chalk stick" or "carrot stick" fractures. Epidural hematomas have also been identified in these patients after trauma even in the absence of fracture. MRI of the entire spine is often indicated for these patients in the setting of trauma to assess for cord or cauda equina compression as well as subtle osseous injuries [4,5].

Early in the disease, radiography can demonstrate lucency at the ante- and/or posterior vertebral body corners which are known as Romanus lesions. Over time, these sites will become sclerotic and produce the aptly named "shiny corner sign." Progressive syndesmophyte formation from ossification of Sharpey's fibers as well as ligamentous ossification produce characteristic fused "bamboo spine" in the late stage of AS (Fig. 9.1). Ossification of interspinous and supraspinous ligaments in particular produces "dagger sign." CT and MRI are typically utilized to assess the complications related to trauma in AS patients (Fig. 9.2). MRI may be utilized to assess active disease in the form of MRI correlate of Romanus lesions and T2/STIR hyperintense vertebral body corners [3–6]. AS patients can also develop Andersson lesion which is hypothesized to represent a chronic fracture or failed fusion of Sharpey fibers across disc space with subsequent advanced degeneration and pseudoarthrosis. The appearance can be strikingly similar to spondylodiscitis but can be differentiated by identifying pseudoarthrosis across both ante- and posterior elements. Lack of adjacent soft-tissue inflammatory changes would also argue against infectious process [7]. Treatment of AS typically involves NSAIDs and physical therapy. In more symptomatic patients, disease-modifying antirheumatic drugs (DMARDs) such as methotrexate or etanercept may be utilized [1].

## 2. Pigmented villonodular synovitis

Pigmented villonodular synovitis (PVNS) is an uncommon disorder of unclear etiology resulting in villous and/or nodular proliferation of synovial tissue in joints, bursae, and tendon sheaths as well as deposition of hemosiderin with progressive joint destruction, osseous erosion, and loss of function. The disease affects approximately 11 per million in the world, typically presenting between ages of 30−50 without sex predilection. In most cases, the disease is monoarticular, preferentially involving knees and less often the hips; however, any synovial joint can be affected including the facet joints [8,9]. Although benign, PVNS can be locally aggressive and complete excision is required to minimize the risk of recurrence [10]. PVNS may be focal (77% of cases) or diffuse (23% of cases) depending on the extent of structural involvement. Extraarticular involvement particular of tendons has also been described and is commonly referred to as

*Multi-system Imaging Spectrum associated with Neurologic Diseases.* https://doi.org/10.1016/B978-0-323-91795-7.00002-6
**109**

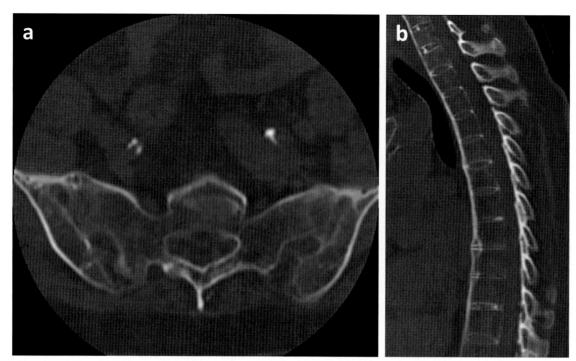

**FIGURE 9.1** (A) Left: An axial CT image of this AS patient demonstrates osseous fusion across the sacroiliac joints. (B) Right: A sagittal CT image of the thoracic spine in the same patient demonstrates osseous fusion across multiple disc spaces from flowing syndesmophytes, forming a "bamboo spine" appearance.

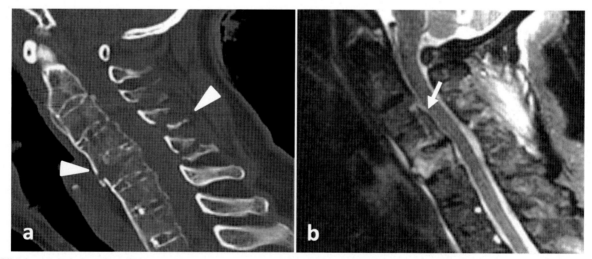

**FIGURE 9.2** (A) Left: A sagittal CT image of an AS patient with ground-level fall and displaced cervical spine fractures of the C5 vertebral body and C3—C5 spinous processes, constituting an unstable spine injury (white arrow heads). (B) Right: A sagittal MRI STIR image of the same patient which demonstrates marrow and paraspinal edema associated with the aforementioned fractures as well as a subtle nondisplaced fracture extending to involve the C3 and C4 vertebral bodies (white arrow). Mild spinal canal narrowing is also present.

"giant cell tumor of the tendon sheath" [11]. Patients present with varying symptoms depending on the joint involved but can include pain, palpable mass or swelling, and less commonly joint dysfunction. Left untreated, joint destruction, osseous erosion, and loss of joint junction ensue. In very rare cases, PVNS can undergo sarcomatous degeneration [8,11].

Radiography is nonspecific and can be normal. CT and MRI may be utilized to image PVNS. CT can demonstrate joint space loss, osseous erosion, and a solitary synovial mass or diffuse synovial hypertrophy demonstrating prominent vascularity. The synovial mass may be hyperdense depending upon the degree of hemosiderin deposition. MRI is the modality of choice and will demonstrate markedly T2 hypointense nodular synovium with blooming artifact on gradient

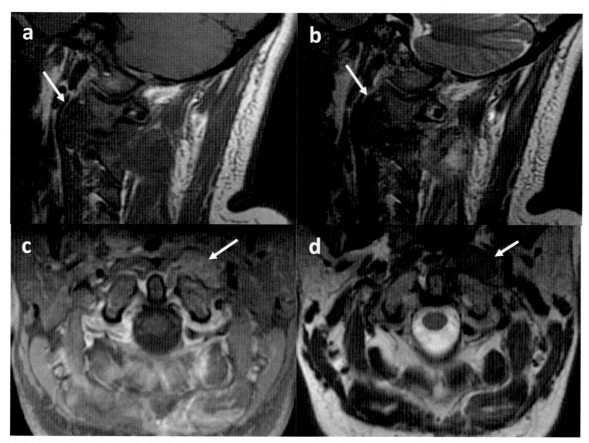

**FIGURE 9.3** (A) Top left (sagittal T1), (B) Top right (sagittal T2), (C) Bottom left (axial T1 postcontrast), and (D) Bottom right (axial T2). MRI images of a patient with PVNS arising from the left C2—C3 facet joint. (A) and (B) demonstrate enlargement of the C2—C3 facet synovium with hypointense signal, more pronounced on the T2 image (white arrows). There is also an associated left dorsal paraspinal collection which demonstrates nodular marginal enhancement as well as heterogeneous internal T2 signal.

echo/susceptibility-weight imaging due to hemosiderin deposition (Fig. 9.3). The nodular synovium may demonstrate variable enhancement. PVNS will also demonstrate F-18 fluorodeoxyglucose (FDG) avidity and uptake on Tc99m-MDP bone scans [9,11]. As previously stated, treatment involves complete surgical resection and possible joint replacement but can be complicated by recurrence. Adjuvant radiation may also be performed on a case-by-case basis. In recurrent cases, amputation may be required [8,10].

## 3. Chordoma

Chordomas are rare midline tumors arising from notochordal remnants which demonstrate slow growth but locally aggressive behavior. Absent treatment, the associated mass effect from its continued growth, and invasion of adjacent structures can result in neurovascular compromise [12,13]. Despite an estimated incidence of 1 case per million per year, these masses account for up to 4% of all malignant bone tumors and up to 40% of malignant sacral tumors. Approximately 50% arise from the sacrum, 35% arise from clivus, and 15% arise from the remainder of spine. Patients span a wide age distribution and are often diagnosed between 40 and 75 years of age with a possible predominance 2:1 male to female [14]. Clival chordoma patients are often diagnosed earlier, between 20 and 55 years old, presumably due to earlier symptom onset [15]. Clinical course is indolent and, if not incidentally found, patients may present with compressive symptoms including pain, neuropathies (such as diplopia from CN VI compression), and incontinence [12—14].

CT demonstrates a midline, well-circumscribed, mildly hyperdense lesion with adjacent osseous erosion and infiltration such that the tumor may contain bone fragments (Fig. 9.4A). On MRI, chordomas appear heterogeneous with T1 hypointense to isointense signal, T2 hyperintense signal, and moderate to marked enhancement (Fig. 9.4B-D). Internal hemorrhage and/or calcification may be present [13—15]. Differential considerations include metastatic disease,

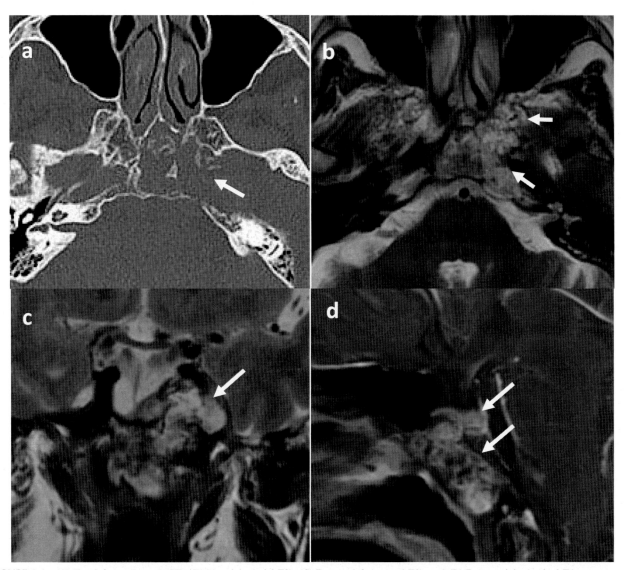

**FIGURE 9.4** (A) Top left (noncontrast CT), (B) Top right (axial T2), (C) Bottom left (coronal T2), and (D) Bottom right (sagittal T1 postcontrast). Patient with clival chordoma. (A) demonstrates a midline clival lytic mass with osseous erosion of the walls of the left carotid canal and left Meckel's cave (white arrows). Bone fragments are present within the mass. (B) and (C) demonstrate the infiltrative nature of the mass extending into aforementioned spaces, cerebellopontine angle, and retro-maxillary soft tissues. The heterogeneous enhancement of this mass is seen in (D) (white arrows).

chondrosarcoma, and ecchordosis physaliphora. The treatment involves aggressive resection and subsequent proton beam radiation, allowing for a 55% survival rate at 10 year. Recurrence is common with 53% rate of recurrence at 5 years and 88% at 10 years [12].

## 4. Extradural meningioma

Primary extradural meningiomas, also termed extracranial meningiomas, are a rare group of meningiomas with atypical location lacking dural attachment. This group constitutes 1%–2% of all meningiomas and may be grouped as "purely calvarial," "purely extracalvarial," and "calvarial with extraosseous extension." The majority of these lesions involve calvarium are also termed "primary intraosseous meningioma." [16,17] In contrast with intradural meningiomas which secondarily involve adjacent bone and soft tissues, these lesions are theorized to arise from meningothelial or arachnoid cap cells which become sequestered within calvarium or adjacent soft tissues, either congenitally or traumatically [18]. As with typical dural meningiomas, there is a bimodal distribution of patients with peaks in the second and fifth decades as well as a slight female predilection. Patients typically present with slowly growing, painless mass without neurologic

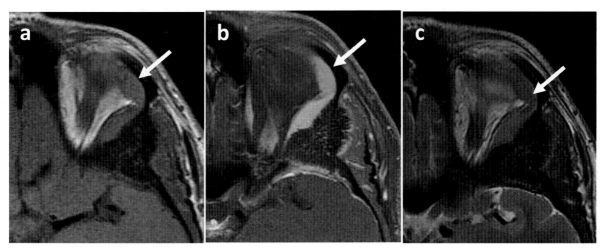

**FIGURE 9.5** (A) Left (axial T1), (B) Middle (axial T1 fat suppressed postcontrast), and (C) Left (axial T2). Patient with primary intraosseous meningioma of the left lateral orbital wall and greater sphenoid wing (white arrows). The mass demonstrates internal calcification and an enhancing soft-tissue component extending into the extraconal soft tissues and epidural space. A small soft-tissue component also underlies the temporalis muscle. Notice that the mass is centered within the bone and the absence of a dural attachment.

symptoms. Depending on the location of mass at the convexity and skull base, patients may also present with longstanding headache, proptosis, visual changes, tinnitus, or other cranial neuropathies [18,19].

On CT, 2/3 of extradural meningiomas are hyperostotic and 1/3 and osteolytic. On MRI, the lesions are mildly T1 hypointense, heterogeneously T2 hyperintense owing to variable calcification, and hyperenhancing (Fig. 9.5) [20]. Differential considerations include osteoma, fibrous dysplasia, osteosarcoma, and metastatic disease. Treatment involves complete or maximal surgical resection with decompression of vital structures and subsequent cranioplasty. Adjuvant radiation therapy may be offered in cases with atypical or malignant histopathology (in up to ¼ of cases) and subtotal resection [16−18].

## 5. Cerebral fat emboli

Fat emboli are an uncommon complication primarily attributed to long bone fractures but also occurring in the setting of fat necrosis which can manifest as fat embolism syndrome (FES). The multiorgan dysfunctions related to the disease process are evident in broad range of symptoms which include fever, tachycardia, respiratory failure, petechial rash, thrombocytopenia, and retinal hemorrhages [21,22]. Cerebral fat emboli are present in 75% of FES patients and cause nonspecific neurologic symptoms such as headache, seizures, and coma [23]. Displaced long bone fractures and orthopedic procedures are commonly cited causes, occurring in 1%−2% of patients. However, cases have been reported in the setting of fat necrosis such as that seen in sickle cell anemia or pancreatitis. The pathophysiology of FES is theorized to be related to microvascular occlusion and/or localized inflammation from the breakdown of embolized fat. Although originally thought to only occur in the setting of right-to-left cardiac shunt lesions, FES can occur in patients without a shunting lesion, and very small fat particles have been demonstrated in animal studies to traverse intact through pulmonary capillary bed [21,22]. Clinical outcomes are often favorable with complete resolution of systemic symptoms. FES-related mortality with appropriate treatment ranges from 7% to 10% [21,22].

MRI demonstrates numerous punctate or confluent T2/FLAIR hyperintense lesions involving periventricular white matter, gray−white interface, basal ganglia, and cerebellum. Characteristically, numerous foci of restricted diffusion involving multiple vascular territories are commonly seen, termed "star field" pattern (Fig. 9.6) [23,24]. Gradient recall echo and susceptibility-weighted imaging demonstrate numerous blooming foci throughout brain due to microhemorrhages [25]. Differential considerations include other causes of star field pattern such as diffuse axonal injury, vasculitis, other embolic materials, and hemorrhagic metastatic disease. FES is treated with supportive care particularly in cases complicated by acute respiratory distress syndrome and early operative fixation of displaced fractures. The administration of steroids to reduce inflammation-mediated sequelae is controversial.

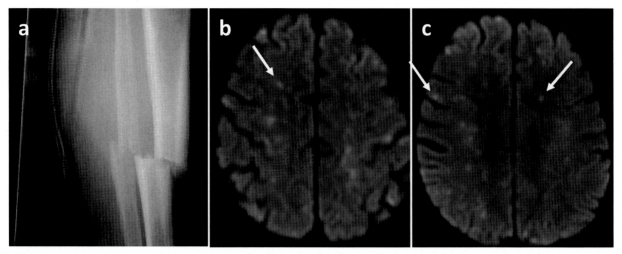

**FIGURE 9.6** (A) Left (radiograph), (B) Middle (DWI-Trace), and (C) Right (DWI-Trace). Patient with mildly displaced tibia and fibula fractures as seen in (A) who presented with altered mentation. The patient underwent MRI which demonstrated multiple bilateral cerebral foci of restricted diffusion (white arrows) involving multiple vascular distributions known as the "star field pattern" which, in the setting of trauma, is consistent with fat embolism.

# 6. Rheumatoid arthritis

Rheumatoid arthritis (RA) is an autoimmune inflammatory arthropathy estimated to affect about 1% of the global population. The disease most commonly arises in 35–50 years of age with 3-to-1 female predominance and increased risk among those with first-degree relatives with RA. The disease has a characteristic symmetric polyarticular distribution particularly involving hands and feet [26,27]. Advanced untreated disease may involve additional areas including dens and atlantoaxial joint with progressive inflammatory infiltration, osseous erosion of dens, and development of retro-dental fibrosis. Dens fracture, ligamentous laxity, and basilar invagination may complicate more severe cases with potentially neurologically devastating consequences depending on the degree of cervico-medullary impingement [28,29]. Improved early recognition of the disease and adoption of DMARD and biologic therapies have significantly reduced the incidence of complications [26]. Diagnosis depends upon a combination of clinical, serologic analysis, and imaging. Patients often present with nonspecific symptoms such as fever, malaise, and myalgias before developing arthropathies which limit activities of daily living. Laboratory tests may demonstrate elevated inflammatory markers such as erythrocyte sedimentation rate and C-reactive protein as well as elevated levels of rheumatoid factor, antinuclear antibody, and other autoantibodies [26,27].

Radiography is often used to assess stability or progression within hands and feet. In patients with disease at cranio-cervical junction, flexion and extension radiography or CT is performed to assess cortical thinning, fracture, ligamentous laxity, dynamic instability, and basilar invagination. MRI can better demonstrate retro-dental fibrosis as well as cervico-medullary compression and associated CNS abnormalities (Fig. 9.7) [28,29]. Treatment with DMARD methotrexate is first line, and biologic agents such as infliximab and etanercept may also be used to augment therapy. Corticosteroids may also be utilized in patients with more refractory disease. Cranio-cervical junction disease may require surgical intervention including decompression and fusion [26–29].

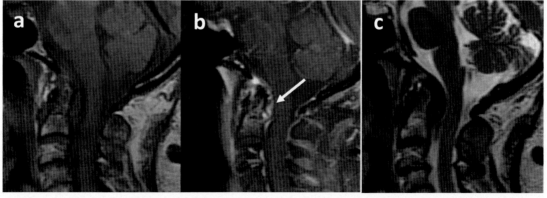

**FIGURE 9.7** (A) Left (sagittal T1), (B) Middle (sagittal T1 fat-suppressed postcontrast), and (C) Right (sagittal T2). Patient with rheumatoid arthritis with T2 hypointense, enhancing retro-dental fibrosis (white arrow) which exerts mass effect upon the cervico-medullary junction. There is suggestion of mild cortical thinning of the underlying odontoid process.

# 7. Osteochondroma

Osteochondroma is a common benign bone tumor appearing as an osseous projection from bone surface with a cartilage cap. They comprise 20%—50% of benign bone tumors and are thought to arise from the periphery of growth plates [30—32]. These lesions are often incidentally found at the metaphysis of long bones, most commonly at the distal femoral metaphysis, and may also more rarely appear in axial skeleton [32,33]. Genetic disorders can predispose to the development of multiple osteochondromas including hereditary multiple exostoses and Trevor's disease (also known as dysplasia epiphysealis hemimelica), the latter of which causes unusual epiphyseal lesions [34]. These lesions are typically pedunculated, but they may also appear sessile as seen particularly in flat bones and spine. These masses are often incidentally found, but patients may also present with painless mass or symptoms related to compression of adjacent structures including the nerves, arteries, and spinal cord. Skull base lesions may also occur leading to cranial nerve compression. In severe cases, lesions may alter biomechanics and cause severe deformity due to mass effect [32—34].

Radiography and CT will demonstrate well-corticated and circumscribed osseous lesions which internally communicate with medullary cavity of the underlying bone (Fig. 9.8). MRI will demonstrate a T2 hyperintense cartilaginous cap measuring up to 1.5 cm in thickness [32,34]. A small number of osteochondromas (approximately 1%—2%) may degenerate into chondrosarcoma as evidenced by thickening of cartilaginous cap and lesion growth after skeletal maturity. Patients with multiple lesions are at increased risk of malignant degeneration by virtue of sheer number of lesions which can warrant MRI screening in this patient population [35—37]. These lesions are often considered "do-not-touch" lesions due to their often-benign behavior. Surgical excision may be performed for symptomatic decompression, correction of biomechanical alteration, and cosmetic reasons [30—34,37].

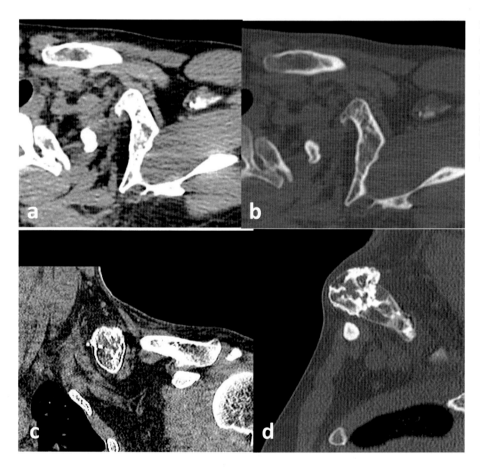

**FIGURE 9.8** (A) Top left (axial soft tissue window CT), (B) Top right (axial bone window CT), (C) Bottom left (coronal CT), and (D) Bottom right (sagittal CT). Patient with left brachial plexopathy. CT images demonstrate an anteriorly and superiorly projecting osteochondroma from the left scapula medial aspect which anteriorly and inferiorly displaces the brachial plexus (A and C). The osteochondroma also tents the skin of the left chest (D).

# References

[1] Zhu W, He X, Cheng K, et al. Ankylosing spondylitis: etiology, pathogenesis, and treatments. Bone Res 2019;7:22.

[2] Dean LE, Jones GT, MacDonald AG, Downham C, Sturrock RD, Macfarlane GJ. Global prevalence of ankylosing spondylitis. Rheumatology (Oxford) 2014;53(4):650—7.

[3] Jang JH, Ward MM, Rucker AN, et al. Ankylosing spondylitis: patterns of radiographic involvement–a re-examination of accepted principles in a cohort of 769 patients. Radiology 2011;258(1):192—8.

[4] Wang YF, Teng MM, Chang CY, Wu HT, Wang ST. Imaging manifestations of spinal fractures in ankylosing spondylitis. AJNR Am J Neuroradiol 2005;26(8):2067—76.

[5] Shah NG, Keraliya A, Nunez DB, et al. Injuries to the rigid spine: what the spine surgeon wants to know. Radiographics 2019;39(2):449—66.

[6] Canella C, Schau B, Ribeiro E, Sbaffi B, Marchiori E. MRI in seronegative spondyloarthritis: imaging features and differential diagnosis in the spine and sacroiliac joints. AJR Am J Roentgenol 2013;200(1):149—57.

[7] Dave BR, Ram H, Krishnan A. Andersson lesion: are we misdiagnosing it? A retrospective study of clinico-radiological features and outcome of short segment fixation. Eur Spine J 2011;20(9):1503—9.

[8] Mendenhall WM, Mendenhall CM, Reith JD, Scarborough MT, Gibbs CP, Mendenhall NP. Pigmented villonodular synovitis. Am J Clin Oncol 2006;29(6):548—50.

[9] Koontz NA, Quigley EP, Witt BL, Sanders RK, Shah LM. Pigmented villonodular synovitis of the cervical spine: case report and review of the literature. BJR Case Rep 2016;2(1):20150264.

[10] Bernthal NM, Ishmael CR, Burke ZDC. Management of pigmented villonodular synovitis (PVNS): an orthopedic surgeon's perspective. Curr Oncol Rep 2020;22(6):63.

[11] Murphey MD, Rhee JH, Lewis RB, Fanburg-Smith JC, Flemming DJ, Walker EA. Pigmented villonodular synovitis: radiologic-pathologic correlation. Radiographics 2008;28(5):1493—518.

[12] Walcott BP, Nahed BV, Mohyeldin A, Coumans JV, Kahle KT, Ferreira MJ. Chordoma: current concepts, management, and future directions. Lancet Oncol 2012;13(2):e69—76.

[13] Soule E, Baig S, Fiester P, et al. Current management and image review of skull base chordoma: what the radiologist needs to know. J Clin Imaging Sci 2021;11:46.

[14] Farsad K, Kattapuram SV, Sacknoff R, Ono J, Nielsen GP. Sacral chordoma. Radiographics 2009;29(5):1525—30.

[15] Santegoeds RGC, Temel Y, Beckervordersandforth JC, Van Overbeeke JJ, Hoeberigs CM. State-of-the-art imaging in human chordoma of the skull base. Curr Radiol Rep 2018;6(5):16.

[16] Elder JB, Atkinson R, Zee CS, Chen TC. Primary intraosseous meningioma. Neurosurg Focus 2007;23(4):E13.

[17] Chen TC. Primary intraosseous meningioma. Neurosurg Clin N Am 2016;27(2):189—93.

[18] Butscheidt S, Ernst M, Rolvien T, et al. Primary intraosseous meningioma: clinical, histological, and differential diagnostic aspects. J Neurosurg 2019:1—10.

[19] Liu Y, Wang H, Shao H, Wang C. Primary extradural meningiomas in head: a report of 19 cases and review of literature. Int J Clin Exp Pathol 2015;8(5):5624—32.

[20] Tokgoz N, Oner YA, Kaymaz M, Ucar M, Yilmaz G, Tali TE. Primary intraosseous meningioma: CT and MRI appearance. AJNR Am J Neuroradiol 2005;26(8):2053—6.

[21] Rothberg DL, Makarewich CA. Fat embolism and fat embolism syndrome. J Am Acad Orthop Surg 2019;27(8):e346—55.

[22] Kwiatt ME, Seamon MJ. Fat embolism syndrome. Int J Crit Illn Inj Sci 2013;3(1):64—8.

[23] Gibbs WN, Opatowsky MJ, Burton EC. AIRP best cases in radiologic-pathologic correlation: cerebral fat embolism syndrome in sickle cell β-thalassemia. Radiographics 2012;32(5):1301—6.

[24] Kuo KH, Pan YJ, Lai YJ, Cheung WK, Chang FC, Jarosz J. Dynamic MR imaging patterns of cerebral fat embolism: a systematic review with illustrative cases. AJNR Am J Neuroradiol 2014;35(6):1052—7.

[25] Giyab O, Balogh B, Bogner P, Gergely O, Tóth A. Microbleeds show a characteristic distribution in cerebral fat embolism. Insights Imaging 2021;12(1):42.

[26] Aletaha D, Smolen JS. Diagnosis and management of rheumatoid arthritis: a review. JAMA 2018;320(13):1360—72.

[27] Smolen JS, Aletaha D, Barton A, et al. Rheumatoid arthritis. Nat Rev Dis Primers 2018;4:18001.

[28] Joaquim AF, Ghizoni E, Tedeschi H, Appenzeller S, Riew KD. Radiological evaluation of cervical spine involvement in rheumatoid arthritis. Neurosurg Focus 2015;38(4):E4.

[29] Gillick JL, Wainwright J, Das K. Rheumatoid arthritis and the cervical spine: a review on the role of surgery. Int J Rheumatol 2015;2015:252456.

[30] Motamedi K, Seeger LL. Benign bone tumors. Radiol Clin North Am 2011;49(6):1115—34 [v].

[31] Lam Y. Bone tumors: benign bone tumors. FP Essent 2020;493:11—21.

[32] Tepelenis K, Papathanakos G, Kitsouli A, et al. Osteochondromas: an updated review of epidemiology, pathogenesis, clinical presentation, radiological features and treatment options. In Vivo 2021;35(2):681—91.

[33] Lotfinia I, Vahedi A, Aeinfar K, Tubbs RS, Vahedi P. Cervical osteochondroma with neurological symptoms: literature review and a case report. Spinal Cord Ser Cases 2017;3:16038.

[34] Murphey MD, Choi JJ, Kransdorf MJ, Flemming DJ, Gannon FH. Imaging of osteochondroma: variants and complications with radiologic-pathologic correlation. Radiographics 2000;20(5):1407—34.

[35] Tong K, Liu H, Wang X, et al. Osteochondroma: review of 431 patients from one medical institution in South China. J Bone Oncol 2017;8:23—9.

[36] Righi A, Pacheco M, Cocchi S, et al. Secondary peripheral chondrosarcoma arising in solitary osteochondroma: variables influencing prognosis and survival. Orphanet J Rare Dis 2022;17(1):74.

[37] Florez B, Mönckeberg J, Castillo G, Beguiristain J. Solitary osteochondroma long-term follow-up. J Pediatr Orthop B 2008;17(2):91—4.

# Chapter 10

# Endocrine system

Raymond Huang, Daniel Phung, Gordon Crews and Nasim Sheikh-Bahaei
*Keck School of Medicine of USC, Los Angeles, CA, United States*

## 1. Background: pituitary adenoma

Pituitary adenomas (PAs) are tumors originating from endocrine cells of the anterior pituitary. Most PAs are slow growing and benign that are classified based on tumor size or cell of origin. Sizes are divided into microadenoma, macroadenoma, and giant tumors. Microadenomas are less than 10 mm, macroadenomas are larger than 10 mm, and giant tumors are larger than 40 mm [1]. About half of all PA are microadenomas with the remaining being macroadenomas [2]. PAs are also classified into three categories based on cell differentiation and expression of transcription factors. Corticotrophs are regulated by the t-box pituitary transcription factor. Somatotrophs, lactotrophs, mammosomatotroph, and thyrotrophs are regulated by pituitary transcription factor 1. Lastly, gonadotrophs are regulated by steroidogenic factor-1 [3]. The majority of PA are frequently prolactinoma (53%), nonfunctioning PA (30%), growth hormone (GH) secreting (12%), and Cushing's disease (4%) [4]. Nonfunctioning PAs are most commonly of gonadotroph cell origin and lack evidence of hormonal secretion except for mild hyperprolactinemia in some cases [5]. They are usually large at the time of diagnosis and present with clinical symptoms such as headache, visual field defects, and hypopituitarism [6]. MRI is currently the recommended imaging modality for diagnosis of pathology related to pituitary gland [7]. It should also be performed in both sagittal and coronal planes with thin 2−3 mm sections and small field of view for optimal results. T1-weighted imaging before and after intravenous contrast as well as T2-weighted should be performed for better detection and evaluation of adenomas [8].

## 2. Prolactinoma

The most common PAs are prolactinomas which comprise 53% of all PA [4] and originate from lactotroph cells of anterior pituitary [9]. The incidence of prolactinoma is 3−5/100,000 year and the prevalence is 50/100,000 [10]. Hyperprolactinemia caused by prolactinoma is considered organic hyperprolactinemia [11], whereas pharmacologic interruption of nigrostriatal dopamine pathway is considered functional hyperprolactinemia [12]. Presentation of hyperprolactinemia is gender specific. Chronic hyperprolactinemia in men typically presents with decreased libido and impotence. In women, hyperprolactinemia presents with galactorrhea and menstrual irregularities [13]. Serum prolactin is used to establish diagnosis of hyperprolactinemia. Prolactin levels greater than 250 μg/L indicate presence of prolactinoma and levels above 500 μg/L are diagnostic of a macroprolactinoma [14].

On MRI, prolactinoma often appears as microadenoma in lateral and posterior portion of anterior pituitary in young women, but usually as a macroadenoma in men and children [15]. Radiographic signs such as cortical thinning and lateral bulging of sella floor are reliable to diagnose prolactinoma combined with other diagnostic studies [16]. Prolactinomas can present with T2 hyperintensity in women while they are more often heterogeneous and lower intensity in men [16]. There have been studies suggest T2 signal intensity may be used as a predictor of prolactinoma response to dopamine agonist therapy [17]. Large prolactinomas may compress optic chiasm and present with bitemporal hemianopsia (Fig. 10.1). Cavernous sinus invasion may also occur which can be assessed by using Knosp classification [18].

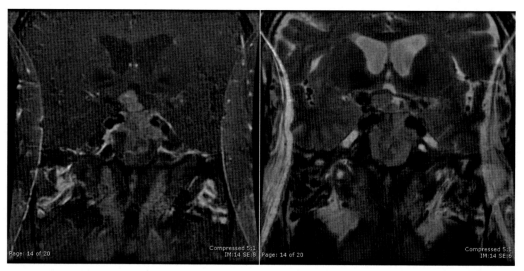

**FIGURE 10.1** Coronal T1WI postcontrast and T2WI demonstrated left-sided macroprolactinoma which is compressing optic chiasm.

# 3. Acromegaly

About two-thirds of PA secrete excess hormones that lead to various endocrine disorders [3]. The most common underlying etiology of acromegaly is a GH-secreting PA arising from somatotrophs and less likely due to ectopic or hypothalamic tumor secretion [19]. The incidence of acromegaly is found to be 11/1,000,000 a year, while the prevalence is 78/1,000,000 [20]. The clinical diagnosis of acromegaly is made based on symptoms related to GH excessive. Serum IGF-1 measurement is used as initial screening for patients with suspected acromegaly [21], while oral glucose suppression test can be used to confirm the diagnosis [22]. Clinical features of acromegaly include, but not limited to, prominence of brow, enlargement of nose and ears, macroglossia, acral enlargement, gigantism, arthralgias and arthritis, and sleep apnea [23]. Some skeletal abnormalities may develop over long periods due to excess GH and IGF-1 which may also appear on radiographs. Patients with acromegaly are at significant risk for vertebral fractures even if the disease is controlled. It is recommended to screen patients with radiographic studies every 2—3 years if they have significant risk factors for osteoporosis or spinal deformities [24]. Acromegaly may have neurologic effects due to skeletal involvement and associated complications as well as due to involvement of optic chiasm and hypothalamus.

As with other PAs, MRI is a neuro-radiological study of choice for acromegaly [25] (Fig. 10.2). Ultrasound may also have a role in providing clues to earlier diagnosis of acromegaly through the detection of a combination of nonspecific

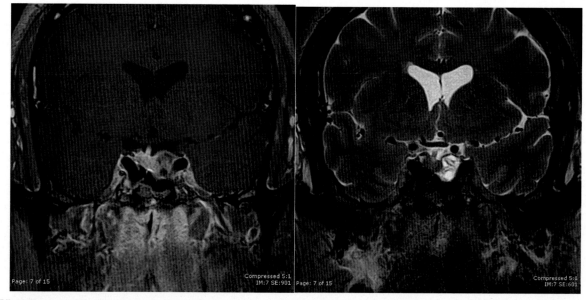

**FIGURE 10.2** Coronal T1WI postcontrast and T2WI demonstrate a left-sided heterogeneously enhancing, T2 hyperintense mass expanding left sella turcica and displacing normal pituitary gland to the right.

multisystem radiographic findings [26]. GH-secreting adenomas are most frequently macroadenomas at the time of diagnosis that exhibit an inverse relationship between the size of adenoma and age of patients [27]. These adenomas are divided into the subtypes of densely granulated and sparsely granulated somatotroph adenoma (DGSA and SGSA) which usually appear as microadenoma in older adults and macroadenoma in younger women, respectively. On T2, DGSA tends to be more hypointense, while SGSA appears more hyperintense [28]. Optic chiasm was compressed more often in women, whereas men had mostly expansion in inferior regions of the sella [27].

## 4. Pituitary hyperplasia

Pituitary hyperplasia (PH) is a nonneoplastic increase in one or more functionally distinct types of pituitary cells that causes pituitary gland to enlarge to greater than 9 mm [29]. Pituitary gland normally undergoes physiological enlargement during periods of hormonal changes such as in pregnancy which can transiently increase up to 10−12 mm during pregnancy and early postpartum [30]. In PH, typically only one cell type becomes hyperplastic and is usually reversible. Some forms of PH may become large enough to mimic neoplastic processes by compressing surrounding structures and may therefore present with neurologic symptoms similar to PA [31]. PH is often underrecognized and epidemiology is unclear; however, the classic cause of PH is end-stage organ insufficiency, most commonly hypothyroidism [32]. In patients with hypothyroidism, pituitary enlargement was found in 37 out of 53 patients (70%) and 84% for patients with TSH ≥100 mIU/mL. 85% of patients treated with levothyroxine showed reversal of PH [33]. Rarely, gonadotroph cell hyperplasia may lead to PH in Turner and Klinefelter syndromes. The associated decrease in serum sex hormone is hypothesized to be the driver of hyperplasia [34].

PH may manifest as symmetric enlargement of the pituitary gland or mimic a PA [35]. It is differentiated from PA by diffuse homogeneous enhancement and isointense signal intensity throughout pituitary gland on both T1WI and T2WI [36] (Fig. 10.3).

## 5. Apoplexy/hemorrhagic adenoma

Pituitary apoplexy is a potentially life-threatening clinical syndrome caused by enlargement of PA due to hemorrhage or infarction [37]. The incidence of pituitary apoplexy is 0.17/100,000 year [38], whereas the prevalence is 6.2/100,000 [19]. Between 2% and 12% of all PAs may experience apoplexy [39]. Presentation of pituitary apoplexy is variable due to extent of hemorrhage, necrosis, and edema. Patients typically present with headaches (>80%), nausea and vomiting (57%) visual disturbances (>50%), photophobia (40%), meningismus (25%), and infrequently fever (16%) [40]. If these symptoms occur within 24−72 h, they are acute symptoms and should be evaluated with MRI. Diagnosis of pituitary apoplexy is confirmed if hemorrhage seen on MRI. Once the patient is medically stable, surgery should be performed to resect hemorrhagic tumor. Patients with subacute symptoms (outside 24−72 h window) can be managed surgically or conservatively with close follow-up MRI monitoring and other testing [41].

CT is useful to quickly rule out subarachnoid hemorrhage and identify pituitary masses in 80% of cases. However, it is only able to identify hemorrhagic components in 20%−30% of cases. On MRI, acute apoplexy appears hypointense on T1W with subtle alternating T1W densities that may suggest apoplexy. Sequential MRI may detect increases in T1W hyperintensity from the periphery to the center of the mass. On T2W, older blood may become heterogeneous and hypointense (Fig. 10.4). T2*W may even detect intratumoral hemorrhage by identifying deposits of hemosiderin [40].

## 6. Thyroid-ophthalmopathy

Thyroid-associated ophthalmopathy (TAO) is an autoimmune disorder that is the most frequent extrathyroidal manifestation of Grave's disease (GD) [42]. The pathogenesis of this disease is not well understood, but it is thought to be caused by activation of orbital fibroblast by autoantibodies associated with GD [43]. The incidence of TAO is 16/100,000 per year for women and 2.9/100,000 per year for men. There is no clear prevalence on TAO, but approximate prevalence is 0.1% −0.3% [44]. Most patients with TAO present with hyperthyroidism (85%), while some are hypothyroid (10%) or euthyroid (5%) [45]. Patients with TAO generally present with lid retraction (90%), proptosis (62%), and restrictive eye movement (43%) [46]. In 40% of patients with GD, both TAO and systemic symptoms of hyperthyroidism present concurrently [47]. Diagnosis of GD is first made by measuring TSH and TSH-R-Ab. Measurement of TSH-R-Ab functionality can be predictive of extrathyroid involvement. Serology combined with ultrasound finding of goiter with larger nodules is reliable for confirmative diagnosis of GD [48]. Patients with TAO may obtain US, CT, or MRI of the orbit to rule out other causes of orbital enlargement. Noncontrast CT is the gold standard for imaging the orbit in TAO as iodinated contrast may exacerbate thyrotoxicosis [49].

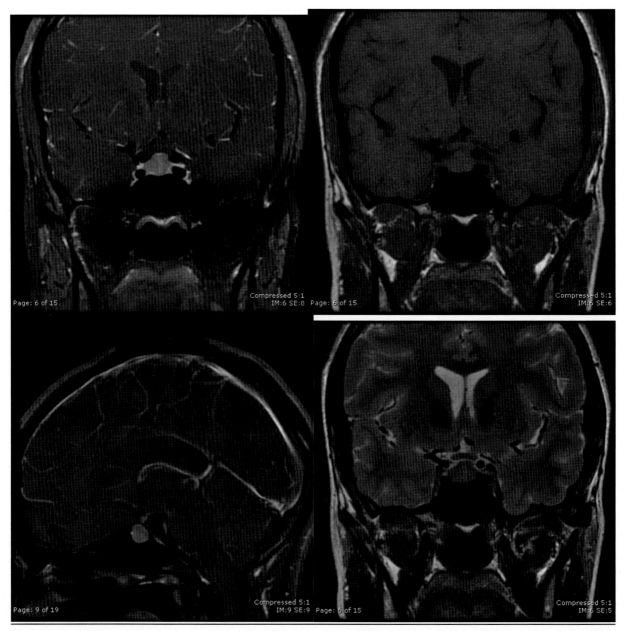

**FIGURE 10.3**    Coronal postcontrast and noncontract T1WI and sagittal postcontrast T1WI and coronal T2WI. There is symmetric enlargement of the pituitary gland which appears homogeneous on all sequences. The infundibulum also appears to be enlarged. The gland extends to the suprasellar cistern without compression of optic chiasm.

On orbit CT, there is typically bilateral involvement of extraocular muscles with sparing of the tendons. The levator palpebrae superior muscle is the first muscle involved typically enlarged to 3 standard deviations from average in about 96% of patients [50]. The most common extraocular muscles involved are the inferior muscle group followed by medial then superior muscles [51]. Enlargement of these extraocular muscles was found to be correlated to clinical presentation. Levator palpebrae superioris enlargement presented with upper lid retraction [52]. If there is concern for optic nerve compression, MRI should be ordered [53] as the presence of optic nerve crowding and intracranial fat prolapse was associated with optic neuropathy [54] (Fig. 10.5).

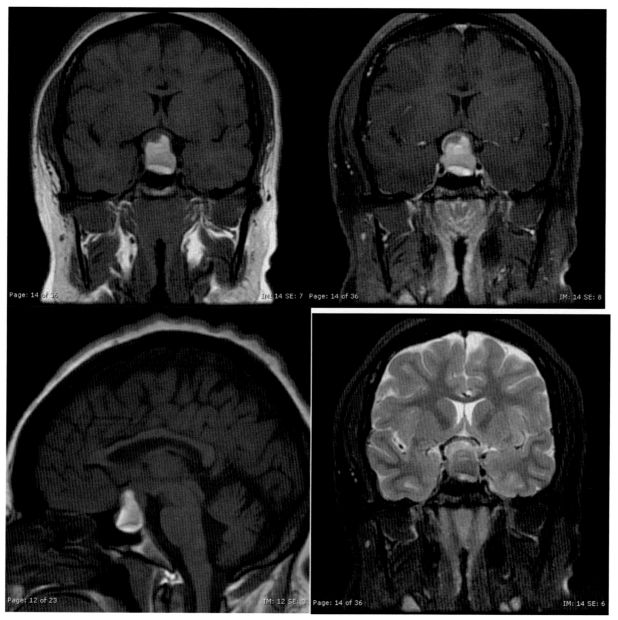

**FIGURE 10.4**  Coronal pre- and postcontrast T1WI, sagittal T1WI, and coronal T2WI demonstrate a large hemorrhagic mass in sella and suprasellar cistern consistent with apoplexy in appropriate clinical setting. The hematoma is compressing optic chiasm and elevating hypothalamus.

## 7. DM and associated neurologic complications

Diabetes mellitus is the seventh leading cause of death in the United States with a prevalence of 10.5% and incidence of 6.9/1000 per year [55]. Diabetes causes increased macrovascular and microvascular aging. The consequences are accelerated atherosclerosis and medial calcification for macrovascular as well as retinopathy and nephropathy for microvascular [56]. Diabetic cerebrovascular diseases are vascular diseases induced by uncontrolled diabetes that result in large and small vessel infarction or hemorrhage [57]. Diabetic patients are 2.9 times more likely to have a stroke compared to nondiabetics [58]. Diabetics also account for 33% of all incidents of ischemic strokes and 26% of all hemorrhagic strokes [59].

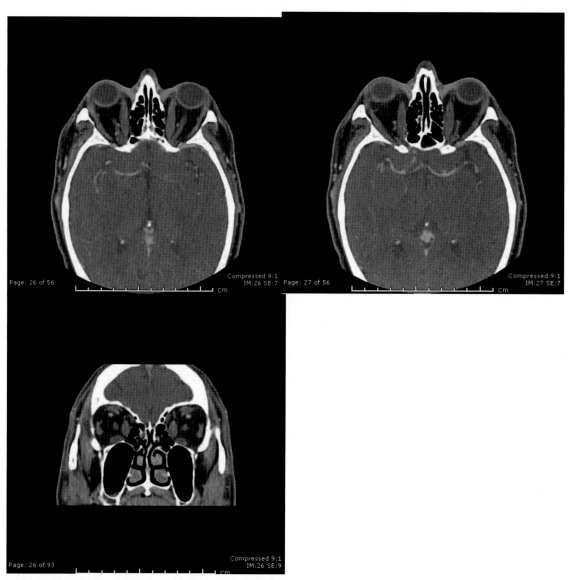

**FIGURE 10.5**  Axial and coronal postcontrast CT demonstrate symmetric bilateral enlargement of extraocular muscles which is most prominently involving medial and inferior rectus muscles.

Diabetics had a higher prevalence for lacunar stroke (32.5%) compared to nondiabetics (10%) [60]. Diabetics typically presented with lower neurological deficits than nondiabetics due to smaller infarct size from lacunar infarct [61].

Diabetics with severe carotid stenosis should be evaluated with carotid US, CT angiography, or MR angiography to determine the appropriate next-step management as these patients are at high risk developing stroke [62]. In patients suspected with stroke, rapid imaging with CT or MRI can be used to differentiate ischemic from intracerebral hemorrhage [63]. MRI T2 FLAIR and diffusion-weighted imaging (DWI) have been shown to be able to estimate onset time of stroke if within 24 h [64]. Despite MRI being more accurate, one of the major issues is the greater time needed to perform compared to CT. MRI may become the primary modality in future as a 6-minute multimodal MRI performed for acute ischemic stroke has similar scan time as conventional CT [65] (Fig. 10.6).

Severe hypoglycemia can also cause encephalopathy associated with radiological changes on MRI. The common findings are T2/FLAIR hyperintensity involving the basal ganglia, posterior limb of internal capsule, hippocampi, and cerebral cortex particularly parieto-occipital regions. These signal changes are also associated with diffusion restriction at earlier stages [66]. Pathophysiology of these changes is not completely understood, which is believed to be secondary to energy failure in hypermetabolic brain regions. The energy shortage results in sodium/potassium pump failure, cellular swelling, and neuronal death [67].

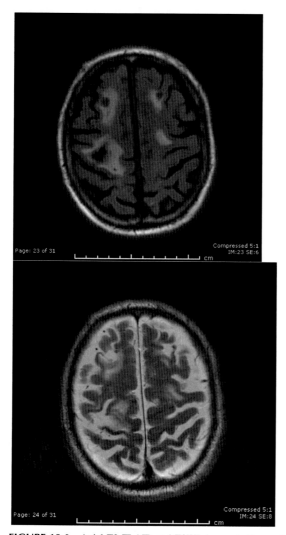

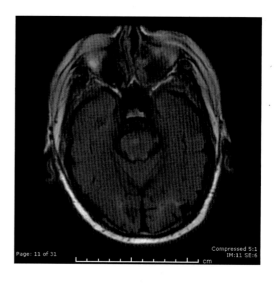

**FIGURE 10.6**    Axial T2 FLAIR and T2WI demonstrating multiple, bilateral, old focal infarcts in different vascular territories in DM patient.

# References

[1] Russ S, Anastasopoulou C, Shafiq I. Pituitary adenoma. Updated 2021 Jul 18. In: StatPearls. Treasure Island (FL): StatPearls Publishing; 2021. Available from: https://www.ncbi.nlm.nih.gov/books/NBK554451/.

[2] Molitch ME. Diagnosis and treatment of pituitary adenomas: a review. J Am Med Assoc 2017;317(5):516−24. https://doi.org/10.1001/jama.2016.19699.

[3] Inoshita N, Nishioka H. The 2017 WHO classification of pituitary adenoma: overview and comments. Brain Tumor Pathol 2018;35(2):51−6. https://doi.org/10.1007/s10014-018-0314-3.

[4] Daly AF, Beckers A. The epidemiology of pituitary adenomas. Endocrinol Metab Clin North Am 2020;49(3):347−55. https://doi.org/10.1016/j.ecl.2020.04.002.

[5] Ntali G, Wass JA. Epidemiology, clinical presentation and diagnosis of non-functioning pituitary adenomas. Pituitary 2018;21(2):111−8. https://doi.org/10.1007/s11102-018-0869-3.

[6] Greenman Y, Stern N. Non-functioning pituitary adenomas. Best Pract Res Clin Endocrinol Metab 2009;23(5):625−38. https://doi.org/10.1016/j.beem.2009.05.005.

[7] Chaudhary V, Bano S. Imaging of the pituitary: recent advances. Indian J Endocrinol Metab 2011;15(Suppl. 3):S216−23. https://doi.org/10.4103/2230-8210.84871.

[8] Evanson J. Radiology of the pituitary. Updated 2020 Jul 19. In: Feingold KR, Anawalt B, Boyce A, et al., editors. Endotext. South Dartmouth (MA): MDText.com, Inc.; 2000. Available from: https://www.ncbi.nlm.nih.gov/books/NBK279161/.

[9] Horvath E, Kovacs K, Scheithauer BW. Pituitary hyperplasia. Pituitary 1999;1(3−4):169−79. https://doi.org/10.1023/a:1009952930425.

[10] Chanson P, Maiter D. The epidemiology, diagnosis and treatment of prolactinomas: the old and the new. Best Pract Res Clin Endocrinol Metab 2019;33(2):101290. https://doi.org/10.1016/j.beem.2019.101290.

[11] Ferrari C, Rampini P, Benco R, Caldara R, Scarduelli C, Crosignani PG. Functional characterization of hypothalamic hyperprolactinemia. J Clin Endocrinol Metab 1982;55(5):897−901. https://doi.org/10.1210/jcem-55-5-897.

[12] Torre DL, Falorni A. Pharmacological causes of hyperprolactinemia. Ther Clin Risk Manag 2007;3(5):929−51.

[13] Galdiero M, Pivonello R, Grasso LF, Cozzolino A, Colao A. Growth hormone, prolactin, and sexuality. J Endocrinol Invest 2012;35(8):782−94. https://doi.org/10.1007/BF03345805.

[14] Melmed S, Casanueva FF, Hoffman AR, et al. Diagnosis and treatment of hyperprolactinemia: an Endocrine Society clinical practice guideline. J Clin Endocrinol Metab 2011;96(2):273−88. https://doi.org/10.1210/jc.2010-1692.

[15] Shih RY, Schroeder JW, Koeller KK. Primary tumors of the pituitary gland: radiologic-pathologic correlation. Radiographics 2021;41(7):2029−46. https://doi.org/10.1148/rg.2021200203.

[16] Kreutz J, Vroonen L, Cattin F, et al. Intensity of prolactinoma on T2-weighted magnetic resonance imaging: towards another gender difference. Neuroradiology 2015;57(7):679−84. https://doi.org/10.1007/s00234-015-1519-3.

[17] Varlamov EV, Hinojosa-Amaya JM, Fleseriu M. Magnetic resonance imaging in the management of prolactinomas; a review of the evidence. Pituitary 2020;23(1):16−26. https://doi.org/10.1007/s11102-019-01001-6.

[18] Knosp E, Steiner E, Kitz K, Matula C. Pituitary adenomas with invasion of the cavernous sinus space: a magnetic resonance imaging classification compared with surgical findings. Neurosurgery 1993;33(4):610−8. https://doi.org/10.1227/00006123-199310000-00008.

[19] Melmed S, Braunstein GD, Horvath E, Ezrin C, Kovacs K. Pathophysiology of acromegaly. Endocr Rev 1983;4(3):271−90. https://doi.org/10.1210/edrv-4-3-271.

[20] Burton T, Le Nestour E, Neary M, Ludlam WH. Incidence and prevalence of acromegaly in a large US health plan database. Pituitary 2016;19(3):262−7. https://doi.org/10.1007/s11102-015-0701-2.

[21] Katznelson L, Laws Jr ER, Melmed S, et al. Acromegaly: an endocrine society clinical practice guideline. J Clin Endocrinol Metab 2014;99(11):3933−51. https://doi.org/10.1210/jc.2014-2700.

[22] Freda PU, Post KD, Powell JS, Wardlaw SL. Evaluation of disease status with sensitive measures of growth hormone secretion in 60 postoperative patients with acromegaly. J Clin Endocrinol Metab 1998;83(11):3808−16. https://doi.org/10.1210/jcem.83.11.5266.

[23] Vilar L, Vilar CF, Lyra R, Lyra R, Naves LA. Acromegaly: clinical features at diagnosis. Pituitary 2017;20(1):22−32. https://doi.org/10.1007/s11102-016-0772-8.

[24] Anthony JR, Ioachimescu AG. Acromegaly and bone disease. Curr Opin Endocrinol Diabetes Obes 2014;21(6):476−82. https://doi.org/10.1097/MED.0000000000000109.

[25] Chanson P, Salenave S. Acromegaly. Orphanet J Rare Dis 2008;3:17. https://doi.org/10.1186/1750-1172-3-17.

[26] Parolin M, Dassie F, Vettor R, Maffei P. Acromegaly and ultrasound: how, when and why? J Endocrinol Invest 2020;43(3):279−87. https://doi.org/10.1007/s40618-019-01111-9.

[27] Potorac I, Petrossians P, Daly AF, et al. Pituitary MRI characteristics in 297 acromegaly patients based on T2-weighted sequences. Endocr Relat Cancer 2015;22(2):169−77. https://doi.org/10.1530/ERC-14-0305.

[28] Trouillas J, Jaffrain-Rea ML, Vasiljevic A, Raverot G, Roncaroli F, Villa C. How to classify the pituitary Neuroendocrine tumors (PitNETs) in 2020. Cancers 2020;12(2):514. https://doi.org/10.3390/cancers12020514.

[29] Chanson P, Daujat F, Young J, et al. Normal pituitary hypertrophy as a frequent cause of pituitary incidentaloma: a follow-up study. J Clin Endocrinol Metab 2001;86(7):3009−15. https://doi.org/10.1210/jcem.86.7.7649.

[30] Elster AD, Sanders TG, Vines FS, Chen MY. Size and shape of the pituitary gland during pregnancy and post partum: measurement with MR imaging. Radiology 1991;181(2):531−5. https://doi.org/10.1148/radiology.181.2.1924800.

[31] Al-Gahtany M, Horvath E, Kovacs K. Pituitary hyperplasia. Hormones (Basel) 2003;2(3):149−58. https://doi.org/10.14310/horm.2002.1195.

[32] De Sousa SM, Earls P, McCormack AI. Pituitary hyperplasia: case series and literature review of an under-recognised and heterogeneous condition. Endocrinol Diabetes Metab Case Rep 2015;2015:150017. https://doi.org/10.1530/EDM-15-0017.

[33] Khawaja NM, Taher BM, Barham ME, et al. Pituitary enlargement in patients with primary hypothyroidism. Endocr Pract 2006;12(1):29−34. https://doi.org/10.4158/EP.12.1.29.

[34] Jentoft M, Scheithauer BW, Moshkin O, et al. Tumefactive postmenopausal gonadotroph cell hyperplasia. Endocr Pathol 2012;23(2):108−11. https://doi.org/10.1007/s12022-012-9196-9.

[35] Connor SE, Penney CC. MRI in the differential diagnosis of a sellar mass. Clin Radiol 2003;58(1):20−31. https://doi.org/10.1053/crad.2002.1119.

[36] Zhang WH, Zhu HJ, Zhang XW, Lian XL, Dai WX, Feng F, et al. Magnetic resonance imaging findings of pituitary hyperplasia due to primary hypothyroidism. Zhongguo Yi Xue Ke Xue Yuan Xue Bao 2012;34(5):468−73. https://doi.org/10.3881/j.issn.1000-503X.2012.05.006.

[37] Semple PL, Webb MK, de Villiers JC, Laws Jr ER. Pituitary apoplexy. Neurosurgery 2005;56(1):65−73. https://doi.org/10.1227/01.neu.0000144840.55247.38.

[38] Raappana A, Koivukangas J, Ebeling T, Pirilä T. Incidence of pituitary adenomas in Northern Finland in 1992-2007. J Clin Endocrinol Metab 2010;95(9):4268−75. https://doi.org/10.1210/jc.2010-0537.

[39] Fernandez A, Karavitaki N, Wass JA. Prevalence of pituitary adenomas: a community-based, cross-sectional study in Banbury (Oxfordshire, UK). Clin Endocrinol 2010;72(3):377−82. https://doi.org/10.1111/j.1365-2265.2009.03667.x.

[40] Briet C, Salenave S, Bonneville JF, Laws ER, Chanson P. Pituitary apoplexy. Endocr Rev 2015;36(6):622−45. https://doi.org/10.1210/er.2015-1042.

[41] Barkhoudarian G, Kelly DF. Pituitary apoplexy. Neurosurg Clin N Am 2019;30(4):457−63. https://doi.org/10.1016/j.nec.2019.06.001.

[42] Şahlı E, Gündüz K. Thyroid-associated ophthalmopathy. Turk J Ophthalmol 2017;47(2):94−105. https://doi.org/10.4274/tjo.80688.

[43] Shan SJ, Douglas RS. The pathophysiology of thyroid eye disease. J Neuro Ophthalmol 2014;34(2):177−85. https://doi.org/10.1097/WNO.0000000000000132.

[44] Lazarus JH. Epidemiology of Graves' orbitopathy (GO) and relationship with thyroid disease. Best Pract Res Clin Endocrinol Metab 2012;26(3):273−9. https://doi.org/10.1016/j.beem.2011.10.005.

[45] Soeters MR, van Zeijl CJ, Boelen A, et al. Optimal management of Graves orbitopathy: a multidisciplinary approach. Neth J Med 2011;69(7):302−8.

[46] Bartley GB, Fatourechi V, Kadrmas EF, et al. Clinical features of Graves' ophthalmopathy in an incidence cohort. Am J Ophthalmol 1996;121(3):284−90. https://doi.org/10.1016/s0002-9394(14)70276-4.

[47] Wiersinga WM, Smit T, van der Gaag R, Koornneef L. Temporal relationship between onset of Graves' ophthalmopathy and onset of thyroidal Graves' disease. J Endocrinol Invest 1988;11(8):615−9. https://doi.org/10.1007/BF03350193.

[48] Kahaly GJ. Management of Graves thyroidal and extrathyroidal disease: an update. J Clin Endocrinol Metab 2020;105(12):3704−20. https://doi.org/10.1210/clinem/dgaa646.

[49] Weiler DL. Thyroid eye disease: a review. Clin Exp Optom 2017;100(1):20−5. https://doi.org/10.1111/cxo.12472.

[50] Del Porto L, Hinds AM, Raoof N, et al. Superior oblique enlargement in thyroid eye disease. J AAPOS 2019;23(5):252. https://doi.org/10.1016/j.jaapos.2019.04.010.

[51] Nugent RA, Belkin RI, Neigel JM, et al. Graves orbitopathy: correlation of CT and clinical findings. Radiology 1990;177(3):675−82. https://doi.org/10.1148/radiology.177.3.2243967.

[52] Davies MJ, Dolman PJ. Levator muscle enlargement in thyroid eye disease-related upper eyelid retraction. Ophthalmic Plast Reconstr Surg 2017;33(1):35−9. https://doi.org/10.1097/IOP.0000000000000633.

[53] Phelps PO, Williams K. Thyroid eye disease for the primary care physician. Dis Mon 2014;60(6):292−8. https://doi.org/10.1016/j.disamonth.2014.03.010.

[54] Birchall D, Goodall KL, Noble JL, Jackson A. Graves ophthalmopathy: intracranial fat prolapse on CT images as an indicator of optic nerve compression. Radiology 1996;200(1):123−7. https://doi.org/10.1148/radiology.200.1.8657899.

[55] Centers for Disease Control and Prevention. National Diabetes Statistics Report, 2017 estimates of diabetes and its burden in the United States background. U.S. Department of Health and Human Services; 2020. Available from: https://www.cdc.gov/diabetes/pdfs/data/statistics/national-diabetes-statistics-report.pdf.

[56] Creager MA, Lüscher TF, Cosentino F, Beckman JA. Diabetes and vascular disease. Circulation 2003;108(12):1527−32. https://doi.org/10.1161/01.cir.0000091257.27563.32.

[57] Zhou H, Zhang X, Lu J. Progress on diabetic cerebrovascular diseases. Bosn J Basic Med Sci 2014;14(4):185−90. https://doi.org/10.17305/bjbms.2014.4.203.

[58] Air EL, Kissela BM. Diabetes, the metabolic syndrome, and ischemic stroke: epidemiology and possible mechanisms. Diabetes Care 2007;30(12):3131−40. https://doi.org/10.2337/dc06-1537.

[59] Lau LH, Lew J, Borschmann K, Thijs V, Ekinci EI. Prevalence of diabetes and its effects on stroke outcomes: a meta-analysis and literature review. J Diabetes Investig 2019;10(3):780−92. https://doi.org/10.1111/jdi.12932.

[60] Ali R. Pattern of stroke in diabetics and non-diabetics. J Ayub Med Coll Abbottabad 2013;25(1−2):89−92.

[61] Tuttolomondo A, Pinto A, Salemi G, et al. Diabetic and non-diabetic subjects with ischemic stroke: differences, subtype distribution and outcome. Nutr Metab Cardiovasc Dis 2008;18(2):152−7. https://doi.org/10.1016/j.numecd.2007.02.003.

[62] Kleindorfer DO, Towfighi A, Chaturvedi S, Cockroft KM, Gutierrez J, Lombardi-Hill D, et al. Guideline for the prevention of stroke in patients with stroke and transient ischemic attack: a guideline from the American Heart Association/American Stroke Association. Stroke 2021;52:e364−467. https://doi.org/10.1161/STR.0000000000000375.

[63] Phipps MS, Jastreboff AM, Furie K, Kernan WN. The diagnosis and management of cerebrovascular disease in diabetes. Curr Diab Rep 2012;12(3):314−23. https://doi.org/10.1007/s11892-012-0271-x.

[64] Aoki J, Kimura K, Iguchi Y, Shibazaki K, Sakai K, Iwanaga T. FLAIR can estimate the onset time in acute ischemic stroke patients. J Neurol Sci 2010;293(1−2):39−44. https://doi.org/10.1016/j.jns.2010.03.011.

[65] Nael K, Khan R, Choudhary G, et al. Six-minute magnetic resonance imaging protocol for evaluation of acute ischemic stroke: pushing the boundaries. Stroke 2014;45(7):1985−91. https://doi.org/10.1161/STROKEAHA.114.005305.

[66] Kang EG, Jeon SJ, Choi SS, Song CJ, Yu IK. Diffusion MR imaging of hypoglycemic encephalopathy. AJNR Am J Neuroradiol 2010;31(3):559−64. https://doi.org/10.3174/ajnr.A1856.

[67] Lee CY, Liou KC, Chen LA. Serial magnetic resonance imaging changes in hypoglycemic encephalopathy. Acta Neurol Taiwan 2013;22(1):22−5.

# Chapter 11

# Immune system

Gordon Crews, Alexander Lerner, Daniel Phung and Nasim Sheikh-Bahaei
*Keck School of Medicine of USC, Los Angeles, CA, United States*

## 1. Acute disseminated encephalomyelopathy

Acute disseminated encephalomyelopathy (ADEM) is a monophasic inflammatory demyelinating disorder involving the brain and spinal cord with onset of symptoms typically following viral illness or vaccination; however, it is thought that it may also be idiopathic in some patients. Clinical symptoms typically develop rapidly, occurring over the course of 2—3 days in the form of generalized encephalopathy, lethargy, seizures, or coma. However, more focal symptoms such as hemiparesis or cranial neuropathies may also develop, depending on the parts of the brain or spinal cord involved [1]. ADEM is typically seen in children and adolescents under the age of 15; however, it can be seen at any age. The prognosis is good with most patients experiencing complete symptomatic resolution following monophasic illness, although a small percentage of patients develop permanent disability or die from their disease [1,2]. The pathophysiology is thought to be an autoimmune response triggered by a viral infection or vaccination resulting in antigen cross-reactivity and perivenular infiltration of lymphocytes, plasma cells, and monocytes that attack oligodendrocytes and result in damage to the myelin without damaging axons. There is damage to gray matter as well [3]. Diagnosis of ADEM is made based on clinical history, laboratory studies, and imaging findings. MRI (magnetic resonance imaging) is the imaging modality of choice when evaluating patients with concern for ADEM.

Imaging findings include large, deep and juxtacortical white matter lesions that are hyperintense on T2 and FLAIR images. Lesions are typically asymmetric. Similar findings can, less frequently, be seen in the deep gray nuclei of basal ganglia and thalami as well as within brainstem and cerebellum. Brain lesions are not commonly seen at calloso-septal interface which is a differentiating feature between ADEM and multiple sclerosis (MS). These lesions can demonstrate diffusion restriction and contrast enhancement. Larger lesions in brain often show ring- or incomplete ring enhancement which is typical of demyelinating lesions also seen in other diseases such as MS [2,4]. Up to a third of patients with brain lesions related to ADEM have spinal cord lesions as well; isolated spinal cord lesions are extremely rare [2]. Imaging features in the spine demonstrate similar characteristics. They demonstrate increased signal on T2 or STIR and are often associated with postcontrast enhancement [2,4]. Lesions may involve both gray and white matter of cord and are typically longer segment lesions (greater than two vertebral body lengths) [5]. Given ADEM typically has a monophasic course of disease, white matter lesions of different ages should not be seen, and multiple lesions may be seen enhancing at the same time [2] (Figs. 11.1 and 11.2).

## 2. Multiple sclerosis

MS is an inflammatory/autoimmune demyelinating disease that affects the central nervous system. Clinical symptoms are nonspecific and can include motor, sensory, and/or autonomic dysfunction depending on specific CNS anatomy involved or affected. The disease process involves a cell-mediated response that attacks myelin components within brain and/or spinal cord and results in loss of oligodendrocytes. While there is no effect upon adjacent axons in acute setting, axonal degeneration is seen in later stages of the disease. In the brain, lesions are typically within the periventricular, deep, or juxtacortical white matter and are seen perpendicular to ventricles/ependymal surface, as this is the orientation of venules to the ventricles, and pathologic analysis of these plaques demonstrates perivenular lymphocytic infiltration [2]. There is often spinal cord involvement as well. Involvement of cervical cord is more common than involvement of thoracic spine.

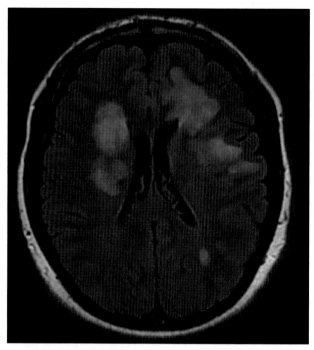

**FIGURE 11.1** Axial FLAIR sequence image demonstrates bilateral, asymmetric hyperintense lesions within deep and juxtacortical white matter in a patient with ADEM.

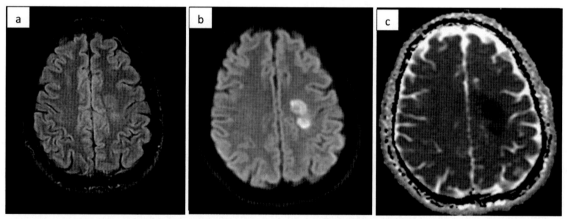

**FIGURE 11.2** A) Patient with ADEM and white matter lesion in left centrum semiovale white matter seen on FLAIR sequence. On diffusion-weighted imaging (B) and ADC map (C) there is associated restricted diffusion which is often seen in active demyelinating lesions associated with ADEM.

In 80% of patients with MS, there is cord involvement; however, only 2%–10% of patients with MS have isolated cord involvement with no brain involvement [6,7].

Diagnosis of MS is typically done with a combination of clinical history and imaging of brain and cervico-thoracic spine with MRI. Per the McDonald criteria, for MS to be diagnosed, there needs to be both geographic and chronologic dissemination of lesions in brain and/or spinal cord [8]. Most commonly, MS has a relapsing-remitting course in which there is rapid onset of symptoms followed by clinical improvement with subsequent relapse. Less common clinical phenotypes include primary progressive disease, in which symptoms rapidly progress following onset of initial symptoms, and the least common, which is progressive relapsing, when patients experience symptomatic exacerbation without complete recovery between attacks [9]. The disease typically manifests between adolescence and 6th decade of life with peak incidence of 35 years of age. There is a predilection for females with female:male ratio of 2:1. Interestingly, the incidence is higher farther away from the equator [9].

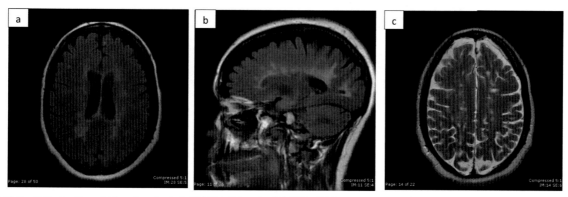

**FIGURE 11.3**  Patient with multiple sclerosis. Axial FLAIR (A) and axial T2WI (B) images show hyperintense periventricular, deep, and juxtacortical white matter lesions consistent with demyelinating lesions. Coronal FLAIR (C) image shows these lesions perpendicular to the lateral ventricles, appearance known as "Dawson's fingers."

MRI is an extremely valuable tool for both diagnosing MS and monitoring progression. In the brain, the most common findings are linear or ovoid lesions in periventricular white matter that demonstrate increased signal on T2/FLAIR (Fig. 11.3). As noted previously, they are typically perpendicular to ependymal surface, as demyelinating plaques are caused by infiltration of perivenular lymphocytes which run perpendicular to ependymal surface. These are often seen at calloso-septal junction and are known as "Dawson's fingers." Calloso-septal location and distribution around temporal horns can help differentiate demyelinating lesions from more nonspecific chronic small vessel disease [6,7]. More nonspecific T2/FLAIR lesions can also be seen in the deep gray nuclei, brain stem, cerebellum, and involving optic nerves. Active demyelinating lesions typically demonstrated associated enhancement on gadolinium T1-postcontrast imaging. This enhancement can appear punctate, linear, or even appear as a mass-like, incomplete ring (known as tumefactive demyelinating lesions). Enhancement is seen for about 1−2 months in duration after onset of symptoms. Diffusion restriction can also be seen associated with active demyelinating lesions [6,7]. In chronic stage, there is evidence of gliosis in demyelinating plaques, and it is associated with volume loss involving both white and gray matter. Thinning of corpus callus is usually present at this stage.

Similar imaging findings can be seen in the spinal cord (Fig. 11.4). As previously mentioned, the cervical cord is more commonly involved compared to the thoracic cord. Demyelinating T2/STIR hyperintense plaques are typically short segment (less than two vertebral bodies in length), oval, and are more often peripheral, although they do not respect gray−white boundary. Postcontrast enhancement is often nodular in appearance when present [6,7]. Diffusion-weighted imaging is not often used for spine imaging.

## 3. Neuromyelitis optica

Neuromyelitis optica (NMO) is an autoimmune demyelinating disease that affects central nervous system. The underlying pathogenesis involves the development of autoantibodies (AQP4-IgG) that target aquaporin-4 (AQP4), which is a membrane channel protein that facilitates the transport of water molecules through cell membrane [10,11]. This membrane channel is highly expressed in various parts of the CNS which include optic nerves, spinal cord, periventricular, hypothalamus, and subependymal regions, as well as brainstem and area postrema [3]. This antigen−antibody complex leads to downregulation of these channels which disrupts water homeostasis and leads to compliment activation, causing damage/death to oligodendrocytes, and vascular damage which leads to symptoms [2]. Because AQP4-IgG has been found in other autoimmune disorders, the term NMO spectrum disorder (NMOSD) is sometimes used. Symptoms of NMO are a direct result of CNS anatomy that is often involved due to the distribution of these membrane channels. Vision loss is a common manifestation due to involvement of optic nerve, and complete acute spinal cord syndrome is another manifestation given involvement of spinal cord. Some patients experience area postrema syndrome of nausea, vomiting, and hiccups related to area postrema involvement. Brainstem involvement can result in multiple other clinical manifestations as well [12]. The average age of onset of NMO symptoms is 42 years of age; there have been cases of pediatric and elderly patients with the disease as well. There is a strong female predilection [2].

Imaging is an integral part of the diagnosis of NMO, especially in patients who are AQP4-IgG-negative. Specifically, it is important in differentiating NMO from other, related differential considerations, namely MS, as the management differs

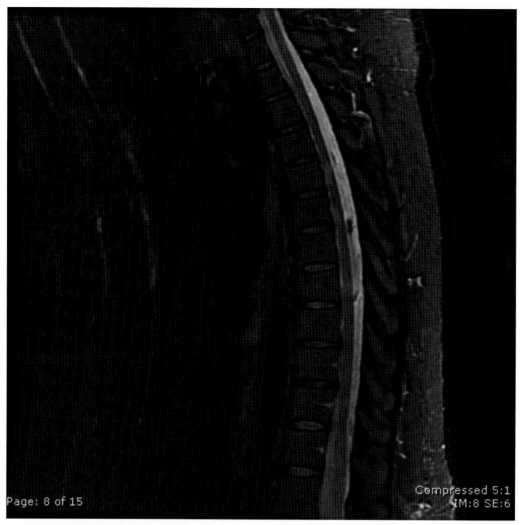

**FIGURE 11.4**   Coronal STIR image through the thoracic spine shows multiple hyperintense demyelinating lesions. They are typical for MS lesions as they are short segment (less than two vertebral body length).

significantly between two diseases. The most common clinical syndrome associated with NMO is optic neuritis. On imaging, acute/subacute phase of NMO associated optic neuritis is characterized by increased caliber/swelling of optic nerves, increased signal on T2 and FLAIR, as well as postcontrast enhancement on T1 postcontrast images (Figs. 11.5 and 11.6). The distribution is typically bilateral, most commonly involving intracranial optic nerves and optic chiasm. The chronic phase will also demonstrate increased signal on T2 and FLAIR sequences; however, optic nerves are typically atrophic in appearance with decreased caliber and do not demonstrate postcontrast enhancement [13]. Spinal cord involvement is also common in NMO. Imaging of acute/subacute disease often shows long-segment (greater than three vertebral bodies in length) lesions which are hyperintense on T2 and STIR with associated patchy or ring-like postcontrast enhancement on T1 postcontrast. Central cord/gray matter is most involved as this is the area surrounding central canal with high AQP4 channel density, typically greater than 50% of transverse surface area of cord is involved. Chronic/remote lesions will not show enhancement, and there is typically increased signal on T2/STIR with areas of decreased caliber [13,14].

MRI findings in the brain include confluent areas of increased signal on T2 and FLAIR images in periventricular white matter and ependymal surfaces of corpus callosum, thalamus (surrounding the third ventricle), hypothalamus, peri-aqueductal region in midbrain, and area postrema. In acute/subacute setting, there can be associated enhancement on T1 contrast-enhanced images. Typically, the enhancement is patchy, although confluent periventricular lesions may demonstrate linear enhancement [13,14]. As previously stated, differentiating NMO from MS is important as the management is different. There is extensive imaging overlap between two entities, and sometimes differentiating the two based on imaging

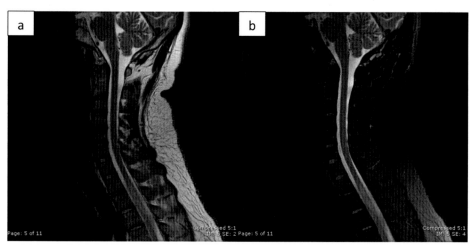

**FIGURE 11.5**   Transverse myelitis in patients with NMO. T2-weighted sequence (A) and STIR sequence (B) demonstrate a long segment lesion (spanning three vertebral bodies) in the upper thoracic cord, which is characterized by expansile, hyperintense signal.

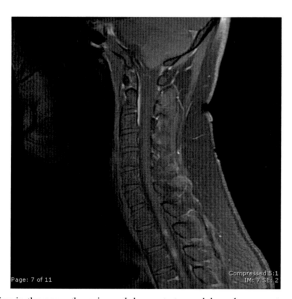

**FIGURE 11.6**   This lesion in the upper thoracic cord demonstrates nodular enhancement on T1 postcontrast images.

alone can be difficult. However, there are some differentiating features that can help in differentiating them. MS typically demonstrates shorter segment cord lesions (less than two vertebral bodies in length) as opposed to the longer segment lesions seen in NMO. MS cord lesions also tend to be more peripheral as opposed to central. Optic nerve involvement in MS is more often unilateral and short segment. Brain lesions have more overlap, although one thing that can sometimes be helpful if ring or ring-like tumefactive lesions are seen, as these are more common in MS and not often seen in NMO [2,14]. Some cases with NMOSD may not have AQP4 antibody but instead they have anti-MOG IgG antibodies. These cases are usually younger compared to typical NMOSD, and in adults, their symptoms are less severe.

## 4. Autoimmune encephalitis

Autoimmune encephalitis is a broad term and encompasses a category of diseases which are characterized by an immune-mediated response established against antigens within brain. Although there are numerous and diverse antibody targets associated with these heterogeneous disease entities, there is a predilection in many of them for specific parts of the brain (e.g., the limbic system and cerebellum) [15]. This often results in overlapping clinical symptoms and imaging patterns among these diseases [16]. Autoimmune encephalitis can be divided into two categories: paraneoplastic and nonparaneoplastic. In paraneoplastic autoimmune encephalitis, immune system develops antibodies to antigen within normal CNS cells that are shared with some type of malignant cells [1]. These paraneoplastic syndromes have been seen associated with numerous types of malignancy. Some of the more classic and common examples include small cell lung cancer (anti-Hu antibodies) and testicular germ cell tumors (anti-Ta antibodies). In nonneoplastic autoimmune encephalitis, the body develops antibodies to native antigens within CNS without an inciting malignancy [16,17]. Antigens involved in paraneoplastic autoimmune encephalitis tend to be intracellular (group 1 antibodies), and those involved in nonneoplastic autoimmune encephalitis tend to be cell-surface antigens (group 2 antibodies), although this is not always the case. This has implications for treatment and outcomes; namely, that treatment of etiologies mediated by group 2 antibodies is often more successful and that they tend to have better outcomes, albeit this is obscured by the fact that there is often no associated underlying malignancy [15]. There are multiple specific antibody-associated syndromes that fall under each category. Signs and symptoms of this group of diseases are a result of the specific anatomic areas targeted by immune response which can lead to variable presentations. Many of the antibodies associated with these diseases have a predilection for antigens within limbic system and/or cerebellum. Presenting symptoms often include memory loss, cognitive dysfunction, seizures, ataxia, dysarthria, or nystagmus [18]. Because of overlap in the location of targeted antigens in many of these district entities, there is much overlap in presenting symptoms. Recognition of imaging patterns that could be suggestive of autoimmune encephalitis is important in leading clinicians down appropriate diagnostic pathway. Age of onset can depend on the typical age that is common for these associated malignancies. Non-neoplastic—associated autoimmune encephalitis typically affects younger people with a slight female predilection.

Imaging is not entirely sensitive, and positive findings on initial imaging are only seen 60%—80% of the time [17]. Contrast-enhanced MRI is the imaging modality of choice for evaluation, although much of the time, it is often performed to rule out other etiologies. Although each different type of antibody has its own predilection for specific anatomy it involves, as mentioned previously, many of these different entities involve overlapping anatomy. One of the most common imaging features is abnormally increased T2 and FLAIR signal involving mesial temporal lobes (specifically limbic system including hippocampi and amygdala), as seen in Fig. 11.7. These findings are more commonly bilateral but may be unilateral as well. Other areas that may demonstrate signal abnormality include deep gray nuclei, brainstem, and cerebral cortex where there is sometimes associated cortical thickening. Although not commonly present, patchy enhancement is sometimes seen within the involved areas [18].

Some types of autoimmune encephalitis, typically paraneoplastic etiologies, involve the cerebellum. This is associated with a variety of different antibodies, the most common being anti-Yo. This is referred to as "paraneoplastic cerebellar degeneration." Imaging findings are similar, with abnormally increased T2 and FLAIR signal with or without patchy areas of enhancement. Another characteristic of this entity is accelerated cerebellar atrophy, which is seen on subsequent imaging [15,17,18]. Although many of these findings are not nonspecific for distinct types of autoimmune encephalitis, and they may share some imaging characteristics with other disease entities (e.g., herpes encephalitis, status epilepticus, etc.), this diverse entity should be suggested if the clinical history is appropriate to help guide clinicians to look for evidence of malignancy or antibodies to correlate with patient symptoms.

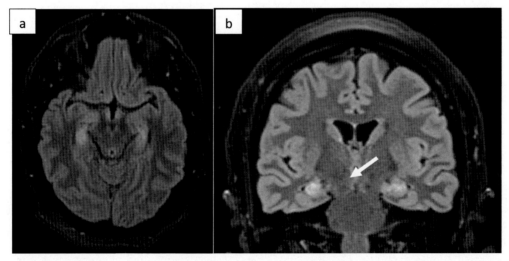

**FIGURE 11.7** Patient with GAD65 autoimmune encephalitis. Axial (A) and coronal (B) FLAIR images demonstrate increased signal within bilateral hippocampi which is an overlapping feature among multiple types of autoimmune encephalitis.

## 5. Neurosarcoidosis

Sarcoidosis is a systemic disease which can affect any part of the body. It is characterized by the formation of noncaseating granulomas. Although it can affect people of any age, people in their third and fourth decade of life are most affected. There is a predilection for people of African American and northern European descent. Women are more commonly affected than men [19,20]. The etiology of sarcoidosis is unknown although there are both genetic and environmental components [21]. The disease process involves formation of noncaseating granulomas because of T-cells responding to unknown antigen [22]. Sarcoid involvement of central nervous system is seen in approximately 25% of people with the disease, only about 5% with neurosarcoidosis are symptomatic from CNS involvement, and about 10% with systemic disease also demonstrate CNS involvement [23]. It is exceedingly rare to have isolated neurosarcoidosis without other systemic evidence of disease. Clinical symptoms of neurosarcoid are nonspecific given that the disease can affect any part of the brain and its surrounding anatomy including meninges, blood vessels, and calvarium. Symptoms can include cranial nerve palsies, headache, seizures, motor, and sensory deficits [21]. Diagnosis of neurosarcoid is based on tissue pathology demonstrating noncaseating granulomas elsewhere in the body in the setting of suggestive radiologic and clinical findings, although when tissue pathology cannot be obtained, suggestive imaging and clinical features can be used to make a probable diagnosis.

MRI is the most sensitive and specific imaging modality when it comes to making findings of neurosarcoidosis. As previously mentioned, sarcoid can involve any part of CNS and its surrounding structures. The most common imaging finding is that of leptomeningeal disease, both in the brain and spine. On MRI, this appears as a nodular leptomeningeal enhancement on postcontrast T1 sequences (Fig. 11.8). There is basilar predilection for leptomeningeal sarcoid involvement which is commonly seen involving cranial nerves. Another imaging manifestation is parenchymal involvement in which nonenhancing lesions that are hyperintense on T2 and FLAIR sequences can be seen in periventricular or juxtacortical white matter or cerebellum (Fig. 11.9). Occasionally, these lesions can also demonstrate postcontrast enhancement. Other findings include nodular or mass-like pachymeningeal thickening and enhancement and even hydrocephalus, due to secondary arachnoid granulation dysfunction [24,25]. Many of these imaging findings are nonspecific and are not pathognomonic for neurosarcoidosis. Leptomeningeal disease can appear like tuberculosis or lymphoma, parenchymal disease can have a similar appearance to lymphoma, metastatic disease, demyelinating processes, or even cerebrovascular disease; and pachymeningeal disease can appear like entities such as idiopathic hypertrophic pachymeningitis. Neurosarcoidosis can also involve spine, although involvement of cervical and thoracic spine is most common. Lesions can be medullary, intradural/extramedullary, or extradural. Imaging features are like findings in the brain as leptomeningeal disease demonstrates nodular pial enhancement, parenchymal lesions are hyperintense on T2/STIR images with occasional

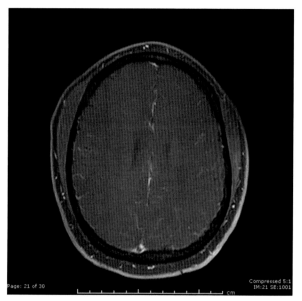

**FIGURE 11.8** The most common imaging manifestation of neurosarcoid is leptomeningeal disease seen here with nodular sulcal/leptomeningeal enhancement as seen on postcontrast T1 image.

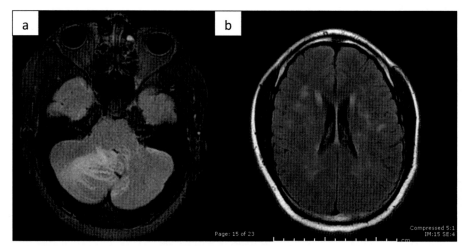

**FIGURE 11.9** Parenchymal involvement of neurosarcoid. (A) and (B) MRI FLAIR images demonstrate periventricular, deep, and juxtacortical lesions as well as a right cerebellar lesion that are hyperintense on T2/FLAIR sequences. Of note, these findings are not specific for neurosarcoid.

enhancement, and dural disease is typically seen as nodular or mass-like thickening and enhancement [24,25]. These findings in spine are nonspecific and can be suggestive of a broad differential diagnosis.

## 6. Systemic lupus erythematosus

Systemic lupus erythematosus (SLE) is an autoimmune disease with numerous clinical manifestations that may affect the central nervous system and result in both neurological and/or psychiatric illness. The term for this is "neuropsychiatric SLE" (NPSLE). Up to 45% of patients with SLE also show evidence of NPSLE [26,27]. Clinical manifestations and symptoms of NPSLE are extremely broad. The American College of Rheumatology have defined 12 central and 7 peripheral presentations which range from mood disorders and cerebrovascular disease (central) to autonomic disorders and cranial neuropathies (peripheral). Clinical diagnosis can be difficult given these nonspecific symptoms/syndrome and the fact that there are no specific or pathognomonic laboratory or imaging findings [27]. The underlying pathophysiology is multifaceted and can be divided into vasculitis/vasculopathy, antiphospholipid syndrome (APS), demyelination, and autoimmune-mediated encephalitis. Many of these processes occur simultaneously to result in clinical manifestations. Autoimmune processes against endothelium of small and medium-sized vessels in brain result in vasculitis/vasculopathy which can result in vascular narrowing and thrombo-embolic disease, infarcts, and intracranial hemorrhage. This can also result in breakdown of blood—brain barrier which allows autoantibodies to reach CSF where it can affect brain. Some of these antibodies can result in demyelination, encephalitis/myelitis, and excitotoxicity [28]. One of the autoantibodies associated with SLE is antiphospholipid (aPL) which can result in APS which is defined as venous or arterial thrombosis in the presence of elevated aPL. This results from activation of endothelium and platelets, creating a prothrombotic environment leading to infarct and early atherosclerosis [28,29].

Although imaging findings of NPSLE are not pathognomonic and are often nonspecific, MRI is the most sensitive modality when it comes to CNS abnormalities associated with this entity. MRI demonstrated findings in about 75% of patients with NPSLE [30]. The most common imaging findings are those of cerebrovascular disease including chronic small vessel ischemic changes and cerebral atrophy which can be seen as increased T2/FLAIR signal within periventricular, deep, and juxtacortical white matter, as well as prominence of ventricles and cerebral sulci, respectively. Infarcts can also be seen in patients with NPSLE, both major vascular distribution infarcts and smaller embolic or lacunar infarcts [28]. Vasculitis is another imaging manifestation that can be appreciated, typically seen as foci of T2/FLAIR hyperintense foci in white matter (Fig. 11.10), which sometimes can be seen in association with punctate postcontrast enhancement (related to blood—brain barrier breakdown) on postcontrast T1WI and punctate susceptibility artifacts on susceptibility-weighted imaging. CT or MR angiogram as well as catheter angiogram can sometimes demonstrate multifocal narrowing of medium-sized vessels [28].

As mentioned previously, demyelinating lesions can also be seen in the setting of NPSLE which can be seen as T2/FLAIR foci within periventricular, deep, and juxtacortical white matter (notice overlapping imaging features with vasculitis) and within brainstem including periaqueductal gray matter and dorsal medulla, which are locations more common in NPSLE than MS. Lesions demonstrating similar imaging characteristics can also be seen in the spinal cord

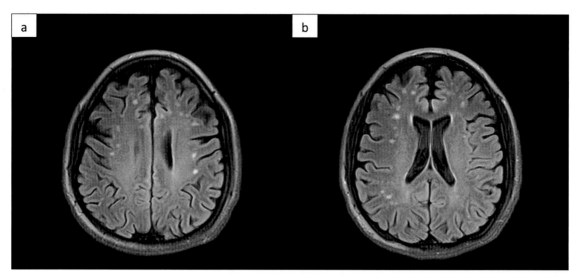

**FIGURE 11.10** SLE-associated vasculitis. (A) and (B) MRI FLAIR demonstrates multiple hyperintense foci within periventricular, deep, and juxta-cortical white matter related to small vessel vasculitis/vasculopathy. Notice that these findings are nonspecific, and there is imaging overlap with multiple other disease processes.

where they are typically long segment (greater than two vertebral bodies in length) which is also not as common as in MS where lesions are typically shorter. Optic neuritis is another manifestation of NPSLE-associated demyelinating syndrome in which optic nerves can demonstrate increased signal on T2/FLAIR [28,29]. Active demyelinating lesions, regardless of location, often demonstrate postcontrast enhancement.

## 7. Bell's palsy

Bell's palsy refers to the clinical condition of unilateral, peripheral facial nerve palsy. It is sometimes referred to as "idiopathic facial nerve palsy," as it is thought of a diagnosis of exclusion and diagnosed after excluding other pathologies that can affect facial nucleus and facial nerve along its course, including ischemic, autoimmune, traumatic, neoplastic, or infectious etiologies. Classically, there is rapid onset of symptoms (within a few hours) with complete resolution within 6–8 weeks (about 2 months) [31]. It results in weakness of the upper and lower parts of the face which clinically differentiates it from a central lesion within the brain. The disease affects people between ages of 15 and 50 years of age, although it can present at any age. Although there are risk factors including diabetes, pregnancy, and hypothyroidism, there is no gender predilection [32]. Understanding the anatomy and course of facial nerve is important in understanding imaging findings which will be discussed below. The facial nerve provides motor innervation to muscles of the face, stapedius, and posterior belly of digastric muscles, although there are additional sensory fibers and parasympathetic fibers that also travel with facial nerve. Facial nucleus lies in dorsal pons, and facial nerve transverses cerebellopontine angle into internal auditory canal where it then enters facial nerve canal of temporal bone to eventually exit stylomastoid foramen and course through parotid gland. Six segments of the nerve include cisternal, canalicular, labyrinthine (from IAC to geniculate ganglion), tympanic, mastoid, and extratemporal segments [33]. The pathophysiology of Bell's palsy is controversial. Even though it is a disease of exclusion and often referred to as "idiopathic," the leading theory is that the etiology is related to reactivation of herpes simplex virus 1 or herpes zoster virus in geniculate ganglion where it spreads to involve adjacent facial nerve and results in viral-mediated inflammation and clinical symptoms [31].

As previously stated, diagnosis of Bell's palsy typically begins with excluding numerous other etiologies, typically with history, physical examination, and imaging. Negative imaging findings (both CT and MRI) of the course of facial nerve including imaging of the brain, internal auditory canals, temporal bone, and face, in the appropriate clinical setting, can suggest diagnosis of Bell's palsy. However, as MR imaging quality has improved, it is now appreciated that there are positive MRI findings to support a diagnosis of Bell's palsy as well [34]. While it has been noted that CT has a role in excluding other etiologies of facial nerve paralysis, it has no role in ruling in the diagnosis of Bell's palsy. Before discussing specific findings on MRI that support a diagnosis of Bell's palsy, it is important to note that there are normal patterns of gadolinium enhancement of the facial nerve on T1 postcontrast images. There can be normal enhancement involving geniculate ganglion and tympanic and mastoid segments [35]. Radiologic–pathological correlation demonstrates these areas of nonpathological enhancement are related to circumneutral facial arteriovenous plexus. These areas of nonpathologic enhancement are not present in every patient and are often asymmetric [35,36]. In Bell's palsy, the most valuable MRI sequence is T1 postcontrast. Typical findings of Bell's palsy are linear, nonnodular enhancement of the distal canalicular through labyrinthine segments as seen in Fig. 11.11. Mass-like enhancement or nodular enhancement can point

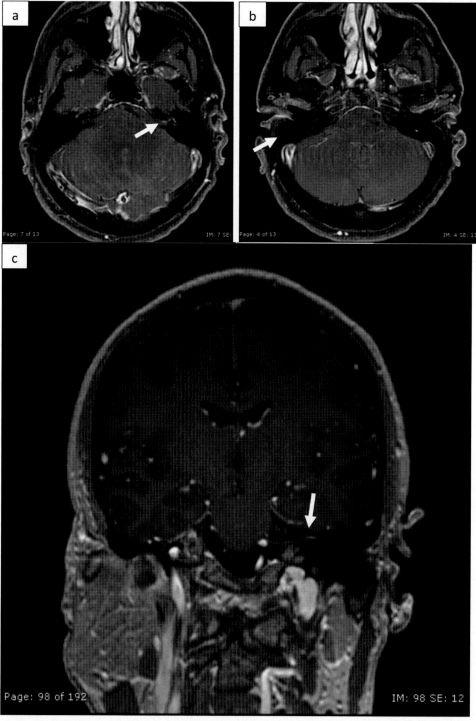

**FIGURE 11.11**  Patient with left-sided facial nerve palsy diagnosed with Bell's Palsy. Multiple axial and coronal postcontrast T1W images through internal auditory canals demonstrate smooth enhancement of left facial nerve involving (A) canalicular segment, (B) tympanic segment, and (C) mastoid segment.

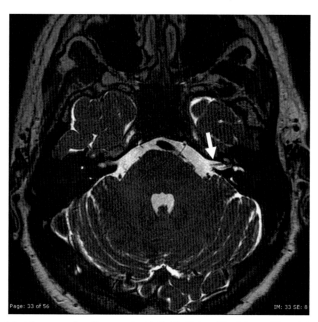

**FIGURE 11.12** Patient with left-sided facial nerve palsy diagnosed with Bell's Palsy. The thin-cut T2-weighted image through the left IAC demonstrates apparent thickening of the facial nerve.

to other etiologies such as a facial nerve sheath tumor or granulomatous disease such as sarcoidosis, respectively. However, even this linear enhancement pattern is not 100% specific for Bell's palsy and should be considered in context of clinical findings. There is some debate as to the sensitivity of MRI findings for Bell's palsy with literature demonstrating a sensitivity of 57%—100%. It is common for imaging findings to be negative [36].

Other reported imaging findings such as linear thickening of the affected facial nerve segments on thin, T2-weighted sequences through the internal auditory canals (Fig. 11.12) are not as sensitive [37].

## 8. CNS vasculitis

CNS vasculitis is a broad group of disease entities characterized by inflammation of intracerebral vasculature which may result in a wide variety of neurological manifestations and symptoms depending on the vessels involved and associated clinical syndromes. Secondary pathology is typically a result of vascular narrowing and/or thrombosis which can result in both reversible and irreversible (infarct) neurological manifestations. Although there is primary, isolated CNS vasculitis known as primary angiitis of the CNS, more often, CNS vasculitis occurs in conjunction with systemic vasculitis syndromes, sometimes even associated with connective tissue disorders [38]. Secondary vasculitis may also occur because of other infectious or neoplastic etiologies or because of certain drug treatments or radiation therapy; however, these secondary causes will not be discussed in detail here. Given CNS vasculitis is a heterogeneous group of disease entities, presenting symptoms are extremely nonspecific and can include fever, fatigue, headache, focal neurologic deficits, psychiatric symptoms, and/or systemic inflammatory signs and symptoms including B-type symptoms such as weight loss or night sweats [39]. Vasculitis is classified by the size or caliber of the affected vessel both within and outside of central nervous system. For example, large vessel vasculitis includes Takayasu arteritis or giant cell arteritis. Medium-sized vessel vasculitis includes Kawasaki disease and polyarteritis nodosa. Small vessel vasculitis includes entities such as granulomatosis with polyangiitis, IgA vasculitis, and microscopic polyangiitis [40]. Although imaging findings associated with distinct types of CNS vasculitis are rarely pathognomonic, the diagnosis of these entities relies heavily on clinical and laboratory findings. Useful laboratory tests include antineutrophil cytoplasmic antibodies which can be seen in association with some small vessel vasculitis. Other, often fewer specific labs include erythrocyte sedimentation rate, rheumatoid factor, C-reactive protein, and cerebrospinal fluid labs such as protein concentration and cell count [41]. However, imaging remains an extremely important piece of the diagnostic puzzle.

Pertinent imaging findings of vasculitis depend on the type of vasculitis and on the size of the vessels involved. Both CT (CT angiogram), and MRI (routine MRI, MR angiography, vessel wall imaging, and perfusion) play a key role in evaluating patients for these diseases. Catheter angiography also plays a significant diagnostic role. Direct imaging findings

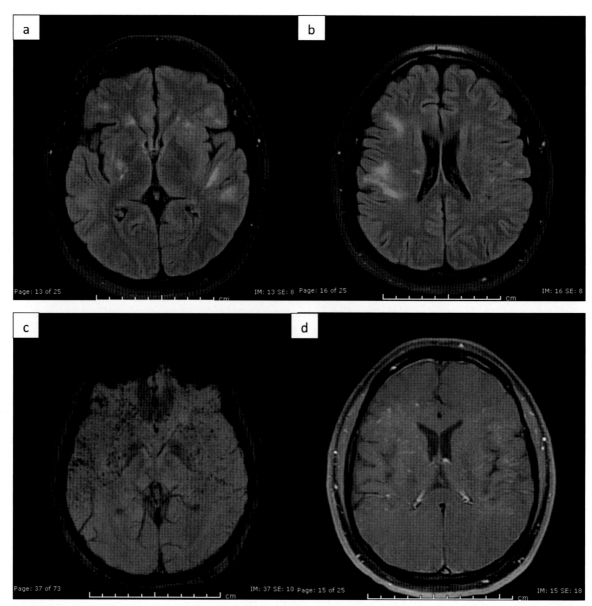

**FIGURE 11.13** Small vessel vasculitis. (A) and (B) FLAIR demonstrates areas of periventricular, juxtacortical, and basal ganglia signal abnormality representing multifocal ischemic changes. (C) Susceptibility-weighted imaging (SWI) shows numerous foci of susceptibility artifact related to mixed age microhemorrhage resulting from friable, inflamed small cerebral arterial vasculature. (D) Postcontrast T1WI showed numerous punctate foci of parenchymal enhancement which results from vessel wall inflammation causing breakdown of blood—brain barrier.

of CNS vasculitis include vessel luminal irregularity/narrowing on vessel imaging and vessel wall thickening and enhancement on vessel wall imaging. Indirect imaging findings which are often less specific but also more frequently encountered include ischemia of different ages within periventricular and juxtacortical white matter as well as within cortex. Punctate parenchymal hemorrhages and subarachnoid hemorrhages of different ages are also a significant finding, related to vessel wall pathology. Some of these findings can be seen in Fig. 11.13, which showed sequela of small vessel vasculitis. Perfusion abnormalities are commonly seen on CT or MR perfusion imaging [40,42].

# References

[1] Hynson JL, Kornberg AJ, Coleman LT, Shield L, Harvey AS, Kean MJ. Clinical and neuroradiologic features of acute disseminated encephalomyelitis in children. Neurology 2001;56(10):1308—12. https://doi.org/10.1212/wnl.56.10.1308.

[2] Sarbu N, Shih RY, Jones RV, Horkayne-Szakaly I, Oleaga L, Smirniotopoulos JG. White matter diseases with radiologic-pathologic correlation. Radiographics 2016;36(5):1426—47. https://doi.org/10.1148/rg.2016160031.

[3] Baum PA, Barkovich AJ, Koch TK, Berg BO. Deep gray matter involvement in children with acute disseminated encephalomyelitis. AJNR Am J Neuroradiol 1994;15(7):1275−83.

[4] Bester M, Petracca M, Inglese M. Neuroimaging of multiple sclerosis, acute disseminated encephalomyelitis, and other demyelinating diseases. Semin Roentgenol 2014;49(1):76−85. https://doi.org/10.1053/j.ro.2013.09.002.

[5] Lee MJ, Aronberg R, Manganaro MS, Ibrahim M, Parmar HA. Diagnostic approach to intrinsic abnormality of spinal cord signal intensity. Radiographics October 2019;39(6):1824−39. https://doi.org/10.1148/rg.2019190021.

[6] Trop I, Bourgouin PM, Lapierre Y, et al. Multiple sclerosis of the spinal cord: diagnosis and follow-up with contrast-enhanced MR and correlation with clinical activity. AJNR Am J Neuroradiol 1998;19(6):1025−33.

[7] Filippi M, Rocca MA. MR imaging of multiple sclerosis. Radiology 2011;259(3):659−81. https://doi.org/10.1148/radiol.11101362.

[8] Hartung HP, Graf J, Aktas O, Mares J, Barnett MH. Diagnosis of multiple sclerosis: revisions of the McDonald criteria 2017 - continuity and change. Curr Opin Neurol 2019;32(3):327−37. https://doi.org/10.1097/WCO.0000000000000699.

[9] Okuda DT. Unanticipated demyelinating pathology of the CNS. Nat Rev Neurol 2009;5(11):591−7. https://doi.org/10.1038/nrneurol.2009.157.

[10] Bennett JL. Finding NMO: the evolving diagnostic criteria of neuromyelitis optica. J Neuro Ophthalmol 2016;36(3):238−45. https://doi.org/10.1097/WNO.0000000000000396.

[11] Wingerchuk DM, Lennon VA, Lucchinetti CF, Pittock SJ, Weinshenker BG. The spectrum of neuromyelitis optica. Lancet Neurol 2007;6(9):805−15. https://doi.org/10.1016/S1474-4422(07)70216-8.

[12] Jasiak-Zatonska M, Kalinowska-Lyszczarz A, Michalak S, Kozubski W. The immunology of neuromyelitis optica-current knowledge, clinical implications, controversies and future perspectives. Int J Mol Sci 2016;17(3):273. https://doi.org/10.3390/ijms17030273.

[13] Barnett Y, Sutton IJ, Ghadiri M, Masters L, Zivadinov R, Barnett MH. Conventional and advanced imaging in neuromyelitis optica. AJNR Am J Neuroradiol 2014;35(8):1458−66. https://doi.org/10.3174/ajnr.A3592.

[14] Dutra BG, da Rocha AJ, Nunes RH, Maia ACM. Neuromyelitis optica spectrum disorders: spectrum of MR imaging findings and their differential diagnosis. Radiographics 2018;38(1):169−93. https://doi.org/10.1148/rg.2018170141.

[15] Kelley BP, Patel SC, Marin HL, Corrigan JJ, Mitsias PD, Griffith B. Autoimmune encephalitis: pathophysiology and imaging review of an overlooked diagnosis. AJNR Am J Neuroradiol 2017;38(6):1070−8. https://doi.org/10.3174/ajnr.A5086.

[16] Degnan AJ, Levy LM. Neuroimaging of rapidly progressive dementias, part 1: neurodegenerative etiologies. AJNR Am J Neuroradiol 2014;35(3):418−23. https://doi.org/10.3174/ajnr.A3454.

[17] Oyanguren B, Sánchez V, González FJ, et al. Limbic encephalitis: a clinical-radiological comparison between herpetic and autoimmune etiologies. Eur J Neurol 2013;20(12):1566−70. https://doi.org/10.1111/ene.12249.

[18] Madhavan AA, Carr CM, Morris PP, et al. Imaging review of paraneoplastic neurologic syndromes. AJNR Am J Neuroradiol 2020;41(12):2176−87. https://doi.org/10.3174/ajnr.A6815.

[19] Pickuth D, Spielmann RP, Heywang-Köbrunner SH. Role of radiology in the diagnosis of neurosarcoidosis. Eur Radiol 2000;10(6):941−4. https://doi.org/10.1007/s003300051042.

[20] Zajicek JP, Scolding NJ, Foster O, et al. Central nervous system sarcoidosis–diagnosis and management. QJM 1999;92(2):103−17. https://doi.org/10.1093/qjmed/92.2.103.

[21] Iannuzzi MC, Maliarik M, Rybicki BA. Nomination of a candidate susceptibility gene in sarcoidosis: the complement receptor 1 gene. Am J Respir Cell Mol Biol 2002;27(1):3−7. https://doi.org/10.1165/ajrcmb.27.1.f243.

[22] Mitchell IC, Turk JL, Mitchell DN. Detection of mycobacterial rRNA in sarcoidosis with liquid-phase hybridisation. Lancet April 25, 1992;339(8800):1015−7. https://doi.org/10.1016/0140-6736(92)90536-c.

[23] Johns CJ, Michele TM. The clinical management of sarcoidosis. A 50-year experience at the Johns Hopkins Hospital. Medicine 1999;78(2):65−111. https://doi.org/10.1097/00005792-199903000-00001.

[24] Smith JK, Matheus MG, Castillo M. Imaging manifestations of neurosarcoidosis. AJR Am J Roentgenol 2004;182(2):289−95. https://doi.org/10.2214/ajr.182.2.1820289.

[25] Ganeshan D, Menias CO, Lubner MG, Pickhardt PJ, Sandrasegaran K, Bhalla S. Sarcoidosis from head to toe: what the radiologist needs to know. Radiographics 2018;38(4):1180−200. https://doi.org/10.1148/rg.2018170157.

[26] Magro-Checa C, Zirkzee EJ, Huizinga TW, Steup-Beekman GM. Management of neuropsychiatric systemic lupus erythematosus: current approaches and future perspectives. Drugs 2016;76(4):459−83. https://doi.org/10.1007/s40265-015-0534-3.

[27] Liang MH, Corzillius M, Bae SC, Lew RA, Fortin PR, Gordon C, Isenberg D, Alarcón GS, Straaton KV, Denburg J, Denburg S. The American College of Rheumatology nomenclature and case definitions for neuropsychiatric lupus syndromes. Arthritis Rheum 1999;42(4):599−608. https://doi.org/10.1002/1529-0131(199904)42:4<599::AID-ANR2>3.0.CO;2-F.

[28] Ota Y, Srinivasan A, Capizzano AA, et al. Central nervous system systemic lupus erythematosus: pathophysiologic, clinical, and imaging features. Radiographics 2022;42(1):212−32. https://doi.org/10.1148/rg.210045.

[29] Cohen D, Rijnink EC, Nabuurs RJ, et al. Brain histopathology in patients with systemic lupus erythematosus: identification of lesions associated with clinical neuropsychiatric lupus syndromes and the role of complement. Rheumatology 2017;56(1):77−86. https://doi.org/10.1093/rheumatology/kew341.

[30] Hanly JG, Kozora E, Beyea SD, Birnbaum J. Review: nervous system disease in systemic lupus erythematosus: current status and future directions. Arthritis Rheumatol 2019;71(1):33−42. https://doi.org/10.1002/art.40591.

[31] Zandian A, Osiro S, Hudson R, et al. The neurologist's dilemma: a comprehensive clinical review of Bell's palsy, with emphasis on current management trends. Med Sci Monit 2014;20:83−90. https://doi.org/10.12659/MSM.889876.

[32] De Diego-Sastre JI, Prim-Espada MP, Fernández-García F. The epidemiology of Bell's palsy. Rev Neurol 2005;41(5):287−90.

[33] Prescott CA. Idiopathic facial nerve palsy (the effect of treatment with steroids). J Laryngol Otol 1988;102(5):403−7. https://doi.org/10.1017/s0022215100105201.

[34] Tien R, Dillon WP, Jackler RK. Contrast-enhanced MR imaging of the facial nerve in 11 patients with Bell's palsy. AJR Am J Roentgenol 1990;155(3):573−9. https://doi.org/10.2214/ajr.155.3.2117359.

[35] Gebarski SS, Telian SA, Niparko JK. Enhancement along the normal facial nerve in the facial canal: MR imaging and anatomic correlation. Radiology May 1992;183(2):391−4. https://doi.org/10.1148/radiology.183.2.1561339.

[36] Hong HS, Yi BH, Cha JG, et al. Enhancement pattern of the normal facial nerve at 3.0 T temporal MRI. Br J Radiol 2010;83(986):118−21. https://doi.org/10.1259/bjr/70067143.

[37] Hector M, Alnadji A, Veillon F, et al. Imaging of facial neuritis using T2-weighted gradient-echo fast imaging employing steady-state acquisition after gadolinium injection. Eur Arch Otorhinolaryngol 2021;278(7):2501−9. https://doi.org/10.1007/s00405-020-06375-z.

[38] Garg A. Vascular brain pathologies. Neuroimaging Clin N Am 2011;21(4):897−926. https://doi.org/10.1016/j.nic.2011.07.007. ix.

[39] Berlit P. Diagnosis and treatment of cerebral vasculitis. Ther Adv Neurol Disord 2010;3(1):29−42. https://doi.org/10.1177/1756285609347123.

[40] Abdel Razek AA, Alvarez H, Bagg S, Refaat S, Castillo M. Imaging spectrum of CNS vasculitis. Radiographics 2014;34(4):873−94. https://doi.org/10.1148/rg.344135028.

[41] Rossi CM, Di Comite G. The clinical spectrum of the neurological involvement in vasculitides. J Neurol Sci 2009;285(1−2):13−21. https://doi.org/10.1016/j.jns.2009.05.017.

[42] Küker W. Cerebral vasculitis: imaging signs revisited. Neuroradiology 2007;49(6):471−9. https://doi.org/10.1007/s00234-007-0223-3.

# Chapter 12

# Lymphatic system

**Jing Yu**

*Department of Radiology, The Affiliated Hospital of Guizhou Medical University, GuiYang, China*

Chapter 12.1

# Sarcoidosis

Sarcoidosis is a multiorgan granulomatous disease of unknown etiology that primarily involves the lungs and the lymphatic system, and characterized by the formation of nonnecrotizing epithelioid granulomas. Sarcoidosis can affect patients of any age, sex, or race, it typically affects adults less than 40 years old, and the incidence peaks in the 3rd decade of life (ages 20–29 years). Extrapulmonary locations can be confined to one organ, or they may be multiply located and diversely combined. The granulomas are of uniform histologic stage, and they have a compact appearance with sharp circumscription from the surrounding lung. Sarcoid granulomas are immune granulomas resulting from a specific cell-mediated response to unidentified antigenic agent/agents.

Bilateral hilar lymph node enlargement, alone or in combination with mediastinal lymph node enlargement, occurs in estimated 95% of patients affected with sarcoidosis. The most common pattern is well-defined, bilateral, symmetric hilar and right paratracheal lymph node enlargement (Fig. 12.1.1). Although sarcoid granulomas arise as micronodular lesions, they may coalesce over time, forming larger lesions (macronodules) (Fig. 12.1.2). Sarcoid granulomas often spread along

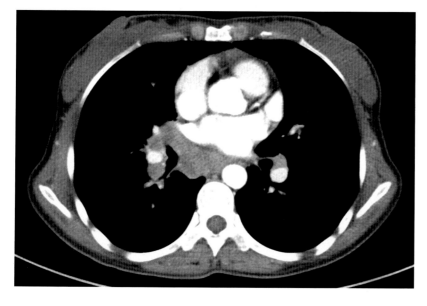

**FIGURE 12.1.1** Radiologic findings of sarcoidosis lymph node lesions. Axial postcontrast CT shows typical bilateral and symmetric hilar and subcarinal lymphadenopathy.

**Multi-system Imaging Spectrum associated with Neurologic Diseases.** https://doi.org/10.1016/B978-0-323-91795-7.00008-7

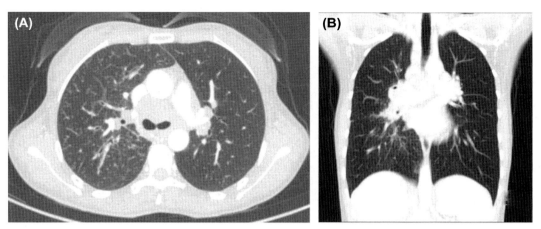

**FIGURE 12.1.2**    Typical manifestations of pulmonary sarcoidosis. Axial (A) and coronal (B) high-resolution CT images shows multiple micronodules with a peri-bronchovascular distribution in the right lung.

the lymphatic vessels, affecting peri-bronchovascular, subpleural, and interlobular septal. In most patients, sarcoid granulomas resolve with time.

Sarcoidosis may involve the spleen and is radiologically similar to hepatic sarcoidosis (Fig. 12.1.3). External thoracothoracic lymphadenopathy usually occurs in the neck, hilar, and intraperitoneal mesenteric regions, but may also occur in atypical sites such as pericardial lymph nodes or intramammary lymph nodes (Fig. 12.1.4). Extrathoracic lymphadenopathies are usually slightly enlarged, rarely confluent, and without calcifications. Nodal enlargement leading to obstruction of the biliary tree and ureters has previously been observed. On MR images, lymphadenopathies usually have a homogeneously increased T2 signal although a central low signal surrounded by a peripheral high signal on T2-weighted images has been observed. Extrathoracic lymph nodes usually show increased FDG activity, but uptake varies widely. Extrathoracic hypermetabolic lymphadenopathies are commonly observed on FDG-PET/CT imaging. Imaging appearances are not specific, and differentiation from metastatic disease, tuberculosis, and malignant lymphoma may be difficult.

**FIGURE 12.1.3**    Splenic involvement in a case of multivisceral sarcoidosis. CT enhancement showed an enlarged spleen with multiple weakly enhanced nodules.

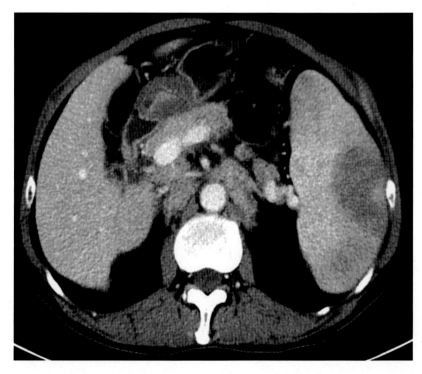

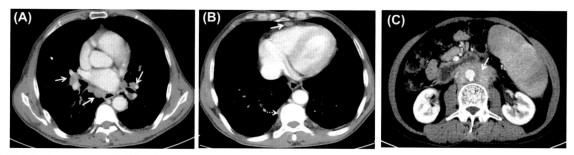

**FIGURE 12.1.4**   Atypical and typical location of lymph node involvement in sarcoidosis. CT enhancement image showing mediastinal and hilar lymph nodes (A arrows), pleural lymph nodes (B dotted arrow), pericardium lymph node(B solid arrow), and retroperitoneal lymph nodes (C arrow).

# Further reading

[1] Reich JM. Mortality of intrathoracic sarcoidosis in referral vs population-based settings: influence of stage, ethnicity, and corticosteroid therapy. Chest 2002;121(1):32−9.

[2] Criado E, Sánchez M, Ramírez J, et al. Pulmonary sarcoidosis: typical and atypical manifestations at high-resolution CT with pathologic correlation. Radiographics October 2010;30(6):1567−86.

[3] Koyama T, Ueda H, Togashi K, Umeoka S, Kataoka M, Nagai S. Radiologic manifestations of sarcoidosis in various organs. Radiographics 2004;24(1):87−104.

[4] Fraioli P, Montemurro L, Castrignano L, et al. Retroperitoneal involvement in sarcoidosis. Sarcoidosis 1990;7:101−5.

[5] Warshauer DM, Lee JK. Imaging manifestations of abdominal sarcoidosis. AJR Am J Roentgenol 2004;182:15−28.

[6] Soussan M, Brillet PY, Mekinian A, et al. Patterns of pulmonary tuberculosis on FDG-PET/CT. Eur J Radiol 2012;81:2872−6.

Chapter 12.2

# Lymphoma

Lymphoma is a tumor that originates in the lymphatic system and is one of the most common types of cancer in the world. Lymphomas are divided into Hodgkin lymphoma, 10%, and non-Hodgkin lymphoma (NHL), 90%. Lymphoma can be localized or diffuse throughout the body. Lymphoma mostly occurs in middle-aged and elderly male patients, and can affect multiple systems throughout the body, including the lymphatic system, central nervous system (CNS), respiratory system, digestive system, musculoskeletal system, genitourinary system, skin, etc. In histological analysis, lymphoma is characterized by single round blue cells with prominent round nuclei and sparse cytoplasm. These characteristics make the lymphoma show a uniform process in imaging. Compared with other tumors, lymphoma's high cell number and high nuclear/cytoplasmic ratio result in lower apparent diffusion coefficient values. This histological property is exploited by DW-MR imaging and leads to limited diffusion. The atypical features of lymphoma include cystic degeneration, necrosis, hemorrhage, and calcification, but they are not common. All these histological features cause lymphoma to lose uniform echo, attenuation, and signal in ultrasound, CT, and magnetic resonance imaging, respectively. Although the angiogenesis of the interstitium is increased in the histological evaluation of aggressive lymphoma, the imaging manifestation of lymphoma is a low-vascular tumor, which is slightly or even not enhanced after intravenous injection of contrast agent.

Primary central nervous system lymphoma (PCNSL) is a rare aggressive high-grade extranodal lymphoma. In PCNSL, lymphoma is confined to the brain parenchyma, meninges, spinal cord, or eyes, and there is no evidence of diseases other than the CNS at the time of initial diagnosis. Most PCNSLs are B-cell type (90% are diffuse large B-cell lymphoma), and a few are T-cell lines. PCNSL is common in patients with weakened immune function, but it can also occur in people with good immune function. Due to the prevalence of human immunodeficiency virus infection and the use of immunosuppressive drugs in transplantation, the incidence of PCNSL is on the rise, from 3.3% before 1978 to 6.6%−15.4% in the early 1990s. However, the decline in morbidity was followed by the use of highly active antiretroviral therapy.

Clinically, PCNSL may mimic other intracranial lesions, such as encephalitis, demyelination, and stroke in imaging. Personality changes, cerebellar signs, headaches, epilepsy, and motor dysfunction may occur. Full symptoms may also occur, but they are more common in T-cell lymphoma. Early diagnosis and treatment can sometimes reduce the irreversible damage of this disease. In addition to primary lymphoma, 10%−15% of patients can also develop secondary systemic lymphoma. Systemic lymphomas are almost always aggressive non-Hodgkin's lymphomas, and the risk of CNS involvement in patients with systemic Hodgkin's disease is very low. The most common manifestation of PCNSL is a single intracranial mass. However, multiple lumps are also common. The typical location of PCNSL is supratentorial, accounting for 70%, and it mostly involves the white matter around the ventricle (Fig. 12.2.1).

Primary pulmonary lymphoma (PPL) is a rare malignant tumor that originates in lymph nodes or extranodal lymphoid tissues (Fig. 12.2.2). The subtypes of primary NHL include mucosa-associated lymphoid tissue (MALT) lymphoma, highly malignant large B-cell lymphoma, hemangiocentric lymphoma, and other rare subtypes, such as intravascular lymphoma. Among them, MALT is the most common, accounting for 60%−80% of all PPL.

The gastrointestinal tract is an important part of the immune system; there are as many lymphocytes in the lymphatic tissue of the intestine as there are lymphocytes in other parts of the body. In patients with diffuse large B-cell lymphoma involving the stomach, CT scans usually show uniform and significant thickening of the stomach wall (i.e., >1 cm), accompanied by delayed tumor enhancement, which can be distinguished from early arterial mucosal enhancement (Fig. 12.2.3). Ulcers and perforations can occur, and there are often larger regional lymph node lesions. However, in the absence of perforations, the fat layer is usually retained (Fig. 12.2.4).

Lymphoma most often involves lymph nodes and immune organs, the latter including thymus, spleen, tonsils, etc. It often manifests as swelling of multiple lymph nodes, and there is a tendency to merge. It can occur in the neck, mediastinum, armpits, groin and abdominal cavity. The imaging findings are similar to those of extranodal lymphoma (Fig. 12.2.5).

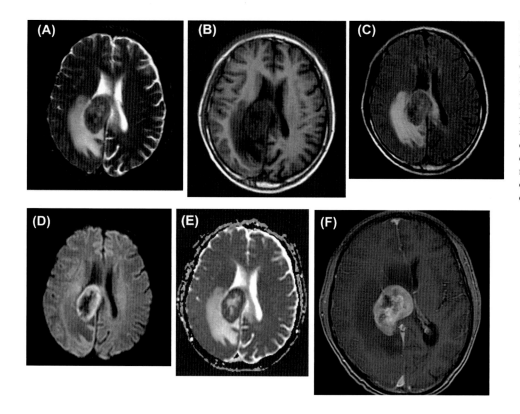

**FIGURE 12.2.1** Intracranial diffuse large B-cell lymphoma. The MRI image shows the mass of the corpus callosum body. On T2-weighted images, mass mostly exhibits low signal, and mass is hypointense to gray matter on T1-weighted imaging. The patchy long T1 long T2 signal area is seen inside, and the necrotic area is considered. A large flaky edema zone can be seen around the mass. The parenchyma of the mass shows high signal on DWI, and low signal on ADC with obvious enhancement.

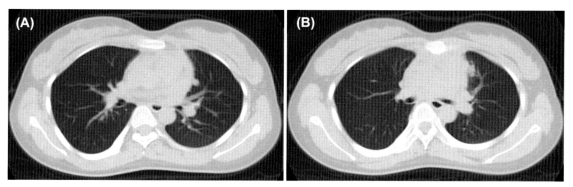

**FIGURE 12.2.2** Multiple quasicircular nodules and masses with clear (A) or unclear (B) boundaries can be seen in the lung interstitium under the pleuras in the lung of patient with lymphoma.

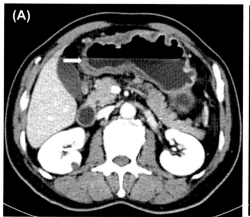
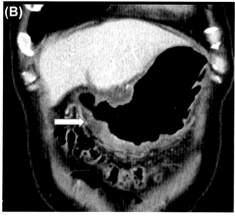

**FIGURE 12.2.3** Diffuse large B cell lymphoma of the stomach. Axial (A) and coronal (B) CT images show marked homogeneous circumferential wall thickening (arrow).

**FIGURE 12.2.4** Diffuse large B-cell lymphoma of the ileum. Axial CT images (A) show a long segment of homogeneous circumferential wall thickening in the distal ileum (star) without evidence of obstruction. Sagittal image (B) demonstrates an ileal-bladder fistula (arrow).

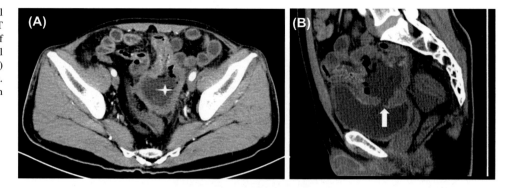

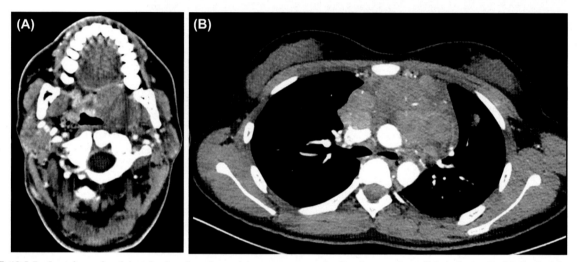

**FIGURE 12.2.5** Lymphoma involving the lymph nodes and immune organs. A: Oropharyngeal tonsil lymphoma. B: Mediastinal lymph node lymphoma.

## Further reading

[1] Mugnaini EN, Ghosh N. Lymphoma. Prim Care December 2016;43(4):661−75.

[2] Koeller KK, Smirniotopoulos JG, Jones RV. Primary central nervous system lymphoma: radiologic-pathologic correlation. Radiographics 1997; 17(6):1497−526.

[3] Mohile NA, Abrey LE. Primary central nervous system lymphoma. Semin Radiat Oncol July 2007;17(3):223−9.

[4] Guzzetta M, Drexler S, Buonocore B, et al. Primary CNS T-cell lymphoma of the spinal cord: case report and literature review. Lab Med 2015;46(2):159−63. Spring.

[5] Diamond C, Taylor TH, Aboumrad T, et al. Changes in acquired immunodeficiency syndrome-related non-Hodgkin lymphoma in the era of highly active antiretroviral therapy: incidence, presentation, treatment, and survival. Cancer January 1, 2006;106(1):128−35.

[6] Tang YZ, Booth TC, Bhogal P, et al. Imaging of primary central nervous system lymphoma. Clin Radiol August 2011;66(8):768−77.

[7] Haque S, Law M, Abrey LE, et al. Imaging of lymphoma of the central nervous system, spine, and orbit. Radiol Clin March 2008;46(2):339−61 [ix].

[8] Haldorsen IS, Espeland A, Larsson EM. Central nervous system lymphoma: characteristic findings on traditional and advanced imaging. AJNR Am J Neuroradiol 2011;32(6):984−92.

[9] Wang Y, Han J, Zhang F, et al. Comparison of radiologic characteristics and pathological presentations of primary pulmonary lymphoma in 22 patients. J Int Med Res April 2020;48(4). 300060519879854.

[10] Manning MA, Somwaru AS, Mehrotra AK, et al. Gastrointestinal lymphoma: radiologic-pathologic correlation. Radiol Clin July 2016;54(4):765−84.

Chapter 12.3

# Leukemias

Leukemia is a hematological malignancy characterized by the production of abnormal white blood cells (leukocytes) in the bone marrow, followed by impaired normal hematopoietic function and decreased cells. Depending on the type of abnormal cells that are produced, leukemias are classified as myeloid (also known as myelogenous or myeloblastic) or lymphoid (also known as lymphocytic or lymphoblastic). In myeloid leukemia, myeloid stem cells (usually mature into red blood cells, platelets, granulocytes, or monocytes) are affected; in lymphoid leukemia, lymphoid stem cells (usually mature into lymphocytes) are affected. Leukemias were initially classified as acute or chronic based on life expectancy, and are now classified as acute or chronic based on cell maturity. In acute leukemia, immature, poorly differentiated cells (blasts) dominate, while in chronic leukemia, mature (but abnormal) cells dominate. In acute leukemia, the onset is usually rapid, while in chronic leukemia, the onset is often slow.

In leukemia, locally aggregated leukemia cells can manifest as mass lesions called granulocytic sarcoma, which can affect multiple organs, including the CNS and the head and neck. Leukemia can affect the meninges (pachymeninges, leptomeninges, or both) (Fig. 12.3.1). Involvement of orbital leukemia can also manifest as focal intraocular or extraocular masses that can affect the eyeballs, lacrimal gland, optic nerve, and different parts of the extraocular muscles. Granulocyte sarcoma can also affect the base of the skull, manifesting as a paravertebral mass with epidural extension and secondary spinal cord compression, although its incidence is much lower than that of intracranial involvement. Granulocyte sarcoma of the spinal cord is extremely rare; however, meningeal leukemia has been reported to also manifest as nerve root enhancement and meningeal enhancement around the spinal cord.

Lymphadenopathy is common in both lymphoid and myeloid leukemia. Systemic lymph nodes can be affected, often involving abdominopelvic l lymph nodes and extramedullary organs. Lymph node biopsy is the gold standard for the diagnosis of lymphadenopathy. However, prior to surgical resection, accurate anatomical distribution information cannot be provided. CT has been widely used since its introduction into clinical practice. Unlike biopsy, Multi-detector computed

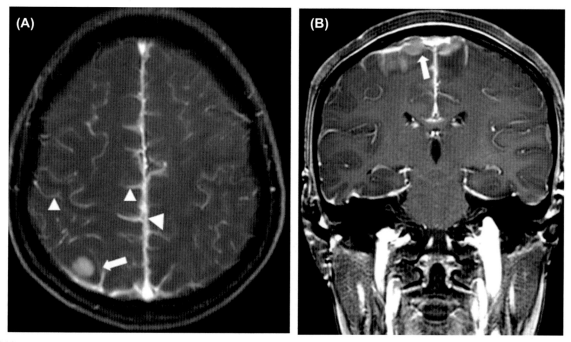

**FIGURE 12.3.1** Leukemic meningitis. Postcontrast axial (A) and coronal (B) T1-weighted images in a 30-year-old woman with acute promyelocytic leukemia demonstrate diffuse dural and leptomeningeal thickening (arrowheads) as well as nodular foci of leptomeningeal enhancement (arrows).

tomography (MDCT) provides complete information about the anatomical distribution of lymph nodes, enhancement patterns, and the appearance of extranodal and extramedullary organs through continuous tomography and multiplanar reconstruction (MPR) (Fig. 12.3.2). Enlarged lymph nodes can fuse, and lymph nodes associated with chronic lymphocytic leukemia fuse more frequently than those associated with acute lymphocytic leukemia and acute myeloid leukemia. Necrosis rarely occurs, so these lymph nodes appear homogeneous on postcontrast CT (Fig. 12.3.3). Necrosis occurs only when lymph nodes are large and the blood supply is blocked; enhanced CT shows low-density areas. Liver and spleen are important reticuloendothelial organs, extranodal and extramedullary organs most commonly involved in leukemia patients. Hepatic sinus, portal tracts, and spleen medulla are the main affected areas. Patients with such involvement have hepatomegaly and splenomegaly due to tumor invasion with or without isolated or multiple low-density lesions (Fig. 12.3.4). Leukemia cell proliferation, infiltration, hemorrhage, malnutrition, necrosis, and secondary infection are pathological basis of low-density disease.

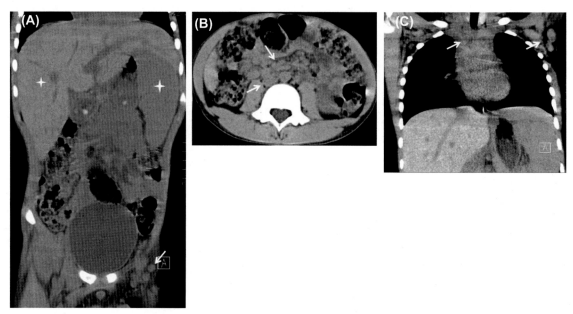

**FIGURE 12.3.2**   An 8-year-old man with acute leukemia on CT images. Hepatosplenomegaly and inguinal lymph node enlargement. B enlarged mesenteric lymph nodes and para-aortic lymph nodes (arrows). C enlarged axillary and mediastinal lymph nodes (arrow).

**FIGURE 12.3.3** A 51-year-old woman with AML on enhanced images. Lymph nodes in the upper para-aortic region conglomerate and show homogenous enhancement (arrow).

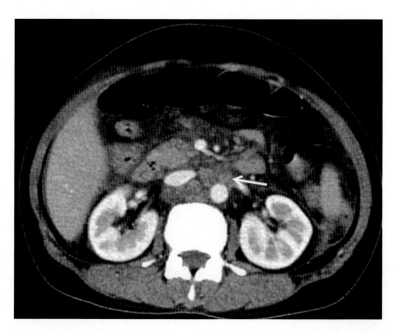

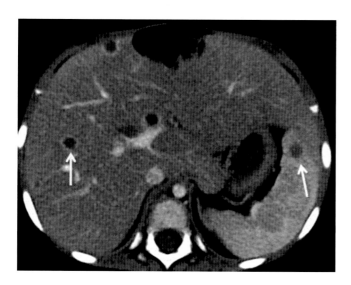

**FIGURE 12.3.4** A 4-year-old woman with ALL on enhanced images. Hepatosplenomegaly and multiple low-density lesions in them are also seen (arrow).

# Further reading

[1] Shroff GS, Truong MT, Carter BW, et al. Leukemic involvement in the thorax. Radiographics 2019;39(1):44−61.

[2] Hakyemez B, Yildirim N, Taskapilioglu O, et al. Intracranial myeloid sarcoma: conventional and advanced MRI findings. Br J Radiol June 2007;80(954):e109−12.

[3] Algharras AA, Mamourian A, Coyne T, et al. Leukostasis in an adult with AML presenting as multiple high attenuation brain masses on CT. J Clin Diagn Res December 2013;7(12):3020−2.

[4] Baikaidi M, Chung SS, Tallman MS, et al. A 75-year-old woman with thoracic spinal cord compression and chloroma (granulocytic sarcoma). Semin Oncol December 2012;39(6):e37−46.

[5] Inamdar KV, Bueso-Ramos CE. Pathology of chronic lymphocytic leukemia: an update. Ann Diagn Pathol October 2007;11(5):363−89.

[6] Matutes E, Attygalle A, Wotherspoon A, et al. Diagnostic issues in chronic lymphocytic leukaemia (CLL). Best Pract Res Clin Haematol March 2010;23(1):3−20.

[7] Walz-Mattmüller R, Horny HP, Ruck P, et al. Incidence and pattern of liver involvement in haematological malignancies. Pathol Res Pract 1998;194(11):781−9.

[8] Lampert IA, Wotherspoon A, Van Noorden S, et al. High expression of CD23 in the proliferation centers of chronic lymphocytic leukemia in lymph nodes and spleen. Hum Pathol June 1999;30(6):648−54.

Chapter 12.4

# POEMS syndrome

POEMS syndrome is a rare paraneoplastic syndrome caused by underlying plasma cell disorders. This acronym was coined by Bardwick in 1980 and refers to several (but not all) features of the syndrome: polyradiculoneuropathy, organomegaly, endocrinopathy, monoclonal plasma cell disorder, and skin changes. The three main points related to this memorable acronym are as follows: 1) not all features in the acronym need to be diagnosed; 2) there are other important features that are not included in the acronym POEMS, including papilledema, extravascular volume overload, sclerotic bone lesions, thrombocytosis/erythrocytosis (P.E.S.T.), elevated vascular endothelial growth factor (VEGF) levels, a predisposition toward thrombosis, and abnormal pulmonary function tests; and 3) a variant of Castleman disease (CD), POEMS syndrome, may be related to a clonal plasma cell disorder. Other less commonly used names for POEMS syndrome are osteosclerotic myeloma, Takatsuki syndrome, or Crow–Fukase syndrome. Because POEMS syndrome involves multiple organs, multiple systems, and multiple disciplines, it is difficult to diagnose and misdiagnosed frequently.

The most serious disabling feature of POEMS is demyelinating neuropathy, which is often misdiagnosed as chronic inflammatory demyelinating polyradiculoneuropathy (CIDP). Asymptomatic pachymeningeal involvement is a very common feature of POEMS syndrome. Thickening of the dura mater is asymptomatic and unrelated to disease duration, but its distribution is unusual, especially in the falx, which may be related to the accumulation of fluid in the meninges. White matter abnormalities and cerebral infarction can also be found in some patients. On MRI, the nerve roots of the brachial plexus and lumbosacral plexus are generally thickened and enhancing, which is difficult to distinguish from CIDP.

Lymph node lesions are one of the characteristics of POEMS syndrome. One of the main diagnostic criteria for POEMS syndrome is CD (also known as angiofollicular hyperplasia or giant lymph node hyperplasia), which has two forms: unicentric and multicentric CD. The diagnosis of CD is very important for the diagnosis, treatment, and prognosis of POEMS. CD often shows clearly demarcated, mildly low-density or iso-density, homogeneous lymph node nodules on nonenhanced CT. It showed moderate and significant enhancement on enhanced CT/MRI (Fig. 12.4.1). MRI shows low-

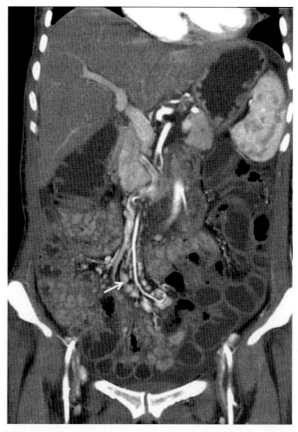

**FIGURE 12.4.1** Castleman disease of POEMS syndrome. Enhanced CT image showed multiple enlarged lymph nodes at the mesenteric root with significant enhancement (arrow).

to-isointensity lesions on T1-weighted images, iso-to-high-intensity lesions on T2-weighted images, and high-intensity lesions on diffusion-weighted images. Calcification and hypertrophy of blood vessels may be valuable diagnostic features.

## Further reading

[1] Khouri J, Nakashima M, Wong S. Update on the diagnosis and treatment of POEMS (polyneuropathy, organomegaly, endocrinopathy, monoclonal gammopathy, and skin changes) syndrome: a review. JAMA Oncol September 1, 2021;7(9):1383−91.

[2] Shi X, Hu S, Yu X, et al. Clinicopathologic analysis of POEMS syndrome and related diseases. Clin Lymphoma, Myeloma & Leukemia January 2015;15(1):e15−21.

[3] Smith C, Saint S, Price R, et al. Clinical problem-solving. Diagnosing one letter at a time. N Engl J Med January 1, 2015;372(1):67−73.

[4] Ziff OJ, Hoskote C, Keddie S, et al. Frequent central nervous system, pachymeningeal and plexus MRI changes in POEMS syndrome. J Neurol May 2019;266(5):1067−72.

[5] Dispenzieri A, Armitage JO, Loe MJ, et al. The clinical spectrum of Castleman's disease. Am J Hematol November 2012;87(11):997−1002.

[6] Li J, Zhou DB. New advances in the diagnosis and treatment of POEMS syndrome. Br J Haematol May 2013;161(3):303−15.

[7] Shi XF, Hu SD, Wu LL, et al. Lymphadenopathy in POEMS syndrome: a correlation between clinical features and imaging findings. Int J Clin Exp Pathol January 1, 2020;13(1):21−5.

[8] Zhao S, Wan Y, Huang Z, et al. Imaging and clinical features of castleman disease. Cancer Imag July 25, 2019;19(1):53.

# Chapter 13

# Integumentary system

**Xia Du**

*Department of Radiology, The Affiliated Hospital of Guizhou Medical University, Guiyang, China*

Chapter 13.1

# Neurofibromatosis

Neurofibromatosis (NF) is a genetic disorder that causes multiple tumors on nerve tissues, including the brain, spinal cord, and peripheral nerves. There are three types in NF: neurofibromatosis type 1 (NF1), neurofibromatosis type 2 (NF2), and schwannomatosis (SWN). NF1 is the most prevalent, accounting for 96% of all cases and characterized by neurofibromas (peripheral nerve tumors) that induce skin changes and bone deformation. NF2 and SWN are rare compared to NF1, occurring in 3% and <1%, respectively. NF2 typically causes hearing loss and vestibular dysfunction, whereas SWN causes intense pain. NF1 is an autosomal dominant genetic disorder characterized by alterations in NF1 gene, resulting in phenotypically heterogeneous systemic manifestations.

NF1 is diagnosed clinically by two or more features including the presence of > six café-au-lait macules, skinfold freckling, Lisch nodules, characteristic lesions of the bone, optic pathway gliomas, neurofibromas of the skin or deep nerve, and a first-degree relative with NF1. Central nervous system (CNS) manifestations of NF1 include neoplasms, learning disabilities, macrocephaly, hydrocephalus, and seizures. NF1 affects a variety of organs and tissues, in the form of neoplasms and nonneoplastic manifestations. Neurofibromas are among the most common manifestations in these patients. They are composed predominantly of a neoplastic Schwann cell, but typically have a variety of soft-tissue and nerve components, including perineurial cells, axons, mast cells, and fibroblasts that likely contribute to tumor growth. Ocular manifestations are also important for the clinical diagnosis of NF1. Lisch nodules (asymptomatic hamartomatous aggregates of melanin-containing cells on surface of iris) occur in almost all patients, and are evaluable through ophthalmologic exam.

Cutaneous findings in NF1 include café-au-lait macules, freckling, neurofibromas, nevus anemicus (NA), juvenile xanthogranuloma (JXG), and so on. The classical café-au-lait macules lesion in NF1 has a round or oval shape with a uniform color of light to dark brown and regular well-demarcated margins (Fig. 13.1.1). Freckling is described as small (1—4 mm), clustered pigmented macules located mostly in axillary and inguinal regions. Typically, these lesions resembling sun-induced freckles appear first in the inguinal region. Neurofibromas skin appears pink to brown, pedicled, or sessile papule nodules. They may be soft or firm in texture ranging from few millimeters to few centimeters located anywhere along the course of peripheral nerves. Blue red macules and pseudoatrophic macules are bizarre variants of cutaneous neurofibroma. These two presentations are less frequent types of neurofibromas and they fulfill the diagnostic criterion of NF1. NA is a congenital, cutaneous anomaly characterized by hypopigmented, well-defined patches with limited vascular supply. The lesions are usually multiple and mostly located on the trunk in varying sizes. JXG is the most common form of non-Langerhans cell histiocytosis characterized by yellowish-pink dome-shaped granulomatous tumors often located on the upper side of the body. It may be solitary or multiple and may start as a pink macule.

NF2 is an autosomal-dominant multiple neoplasia syndrome that results from mutations in the NF2 tumor suppressor gene located on chromosome 22q. Patients are predisposed to development of lesions of the nervous system, eyes, and

Multi-system Imaging Spectrum associated with Neurologic Diseases. https://doi.org/10.1016/B978-0-323-91795-7.00007-5

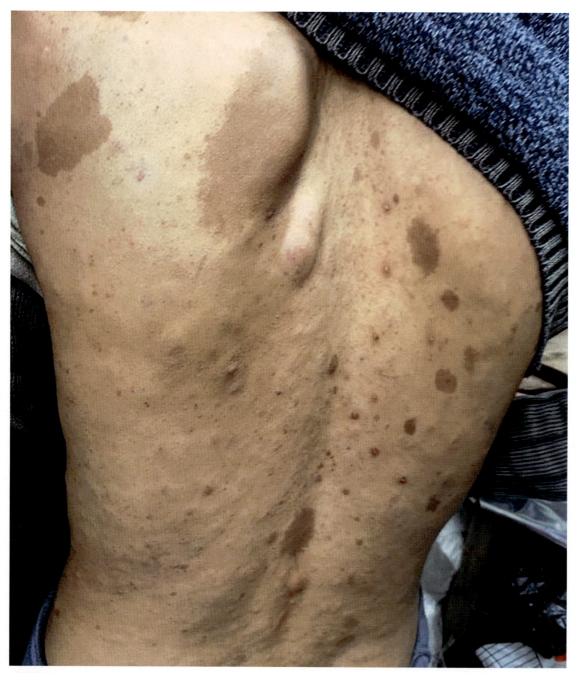

**FIGURE 13.1.1** Patients with neurofibromatosis type 1, multiple café-au-lait macules of varying size are seen on the lower back.

skin. Skin tumors are present in 59%—68% of patients with this disorder and include skin plaques and subcutaneous disorder. Most skin tumors are schwannomas, although neurofibromas or mixed tumors have been occasionally identified. Skin plaques are well-circumscribed, slightly raised, roughened areas that typically are less than 2 cm and display slight hyperpigmentation and hypertrichosis. They are present in 41%—48% of patients and might be hairless, smooth, and soft in patients under 10 years. These tumors are present in 43%—48% of patients and are often painful and sensitive to pressure. Café-au-lait maculae are well-defined flat, hyperpigmented areas of skin that represent a nonspecific finding.

NF2 mainly involves the CNS, and the most prominent feature is bilateral acoustic neuroma, which can be associated with meningioma, ependymoma, and other neurilemmoma of cerebral nerve and spinal nerve. The acoustic neuroma showed a space-occupying lesion in the cerebellopontine angle area, with the internal auditory canal as the center, with enlargement of the internal auditory canal, symmetrical or asymmetric tumor size, slightly high density on CT, low signal

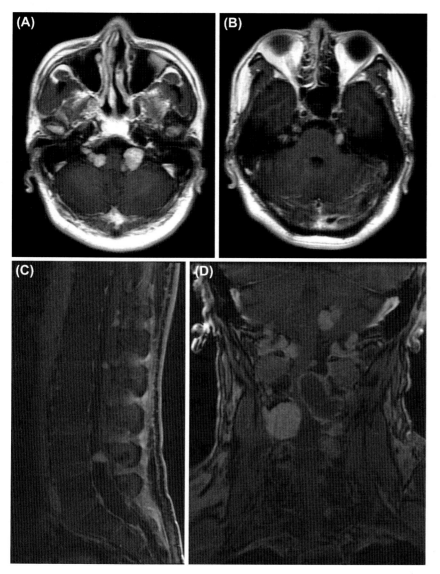

**FIGURE 13.1.2** Patients with neurofibromatosis type 2, male, 52 years old, limb numbness, fatigue and poor exercise for more than 1 year, aggravated for 1 month. (A—B): The enhanced transverse and coronal MRI of the head showed thickening of bilateral auditory nerve and nodular thickening of bilateral trigeminal nerve. (C—D): The sagittal enhanced MRI of lumbar vertebrae and cervical MRI-enhanced coronal images showed multiple abnormal signals in the spinal canal and paraspinal canal, with obvious enhancement, and some lesions showed circular enhancement.

intensity on MRI, iso/slightly high signal on T2WI, low/slightly high signal on FLAIR, long T1 and long T2 signal cyst necrotic area in some tumors, and homogeneous enhancement in solid part, but no enhancement in cystic necrotic area. Meningiomas associated with NF2 are similar to spontaneous meningiomas in location and imaging findings, but they are young and often have multiple meningiomas. Spinal ependymoma is often located in the neck, thoracic spinal cord, or spinal cord cone, and its imaging findings are similar to those of spontaneous ependymoma. Clinically, the symptoms and signs of NF2 are not specific, but its imaging findings have significant features (Fig. 13.1.2).

# Further reading

[1] Gutmann DH, Ferner RE, Listernick RH, Korf BR, Wolters PL, Johnson KJ. Neurofibromatosis type 1. Nat Rev Dis Prim February 23, 2017;3:17004.

[2] Ozarslan B, Russo T, Argenziano G, Santoro C, Piccolo V. Cutaneous findings in neurofibromatosis type 1. Cancers January 26, 2021;13(3):463.

[3] Jacques C, Dietemann JL. Imagerie de la neurofibromatose de type 1. J Neuroradiol 2005;32:180—97.

[4] Nix JS, Blakeley J, Rodriguez FJ. An update on the central nervous system manifestations of neurofibromatosis type 1. Acta Neuropathol April 2020;139(4):625—41.

[5] Asthagiri AR, Parry DM, Butman JA, Kim HJ, Tsilou ET, Zhuang Z, et al. Neurofibromatosis type 2. Lancet June 6, 2009;373(9679):1974—86.

[6] Girard N. Imagerie de la neurofibromatose de type 2. J Neuroradiol 2005;32:198—203.

[7] Tamura R. Current understanding of neurofibromatosis type 1, 2, and schwannomatosis. Int J Mol Sci May 29, 2021;22(11):5850.

[8] Ferner RE. Neurofibromatosis 1 and neurofibromatosis 2: a twenty first century perspective. Lancet Neurol April 2007;6(4):340−51.

[9] Ahlawat S, Blakeley JO, Langmead S, Belzberg AJ, Fayad LM. Current status and recommendations for imaging in neurofibromatosis type 1, neurofibromatosis type 2, and schwannomatosis. Skeletal Radiol February 2020;49(2):199−219.

[10] Ahlawat S, Fayad LM, Khan MS, et al. Current whole-body MRI applications in the neurofibromatoses: NF1, NF2, and schwannomatosis. Neurology August 16, 2016;87(7 Suppl. 1):S31−9.

[11] Kilpatrick TJ, Hjorth RJ, Gonzales MF. A case of neurofibromatosis 2 presenting with a mononeuritis multiplex. J Neurol Neurosurg Psychiatry 1992;55:391−3.

[12] Gijtenbeek JMM, Gabreels-Festen AAWM, Lammens M, Zwarts MJ, Van Engelen BGM. Mononeuropathy multiplex as the initial manifestation of neurofibromatosis type 2. Neurology 2001;56:1766−8.

Chapter 13.2

# Tuberous sclerosis

Tuberous sclerosis (TSC) is an autosomal dominant neurocutaneous syndrome characterized by various abnormalities, including benign hamartomatous tumors in multiple organs. TSC was first described by von Recklinghausen and then named by Bourneville for its gross neuropathological appearance. The incidence of the disease is about 1 in 20,000−50,000, most common in children, male more than female 2−3 times, which can involve the brain, skin, kidney, heart and other body organs and tissues. The typical clinical features were epilepsy, adenoma sebaceous of the face, and mental retardation. Epilepsy occurs in more than 70%−80% of patients with TSC. TSC is an autosomal dominant disorder that affects multiple organ systems and is caused by loss-of-function mutations in one of two genes: TSC1 or TSC2. TSC1 is located on chromosome 9q34 and encodes the protein hamartin. TSC2 is found on chromosome 16p13 and encodes protein tuberin. Hamartin and tuberin interact to form heterodimers, which inhibit mammalian target of rapamycin (mTOR) −signaling cascade, which is responsible for regulating cell growth and differentiation. The absence of either hamartin or tuberin leads to a loss of inhibition of mTOR and therefore production of multisystemic hamartomatous tumors of TSC. Mutations of TSC2 are much more frequent than mutations of TSC1 and are associated with more severe disease. In general, this abnormal proliferation is limited and does not lead to malignant transformation.

Skin manifestations develop in almost all individuals with TSC. These include facial angiofibromas, ungual fibromas, fibrous cephalic plaques, shagreen patches, and focal hypopigmentation changes. Hypomelanotic macules, also known as ash leaf spots, are the earliest dermatologic changes and are present at birth and usually linear and oval in shape on the trunk and limbs, ranging from a few millimeters to a few centimeters. More than 90% of the children had sebaceous adenoma by age four. Sebaceous adenoma is mainly composed of sebaceous glands, connective tissue and blood vessels, appear as reddish papules, and typically present in a butterfly or malar distribution in the nasolabial sulcus chin and buccal (Fig. 13.2.1). After the age of 10, children may

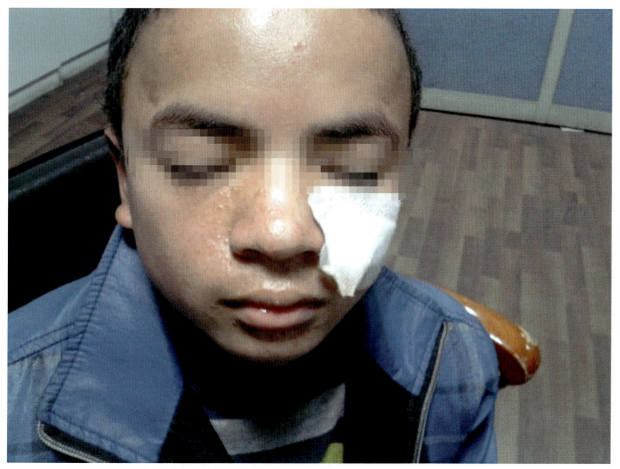

**FIGURE 13.2.1**   Patient with tuberous sclerosis, multiple facial sebaceous adenomas are seen on the face.

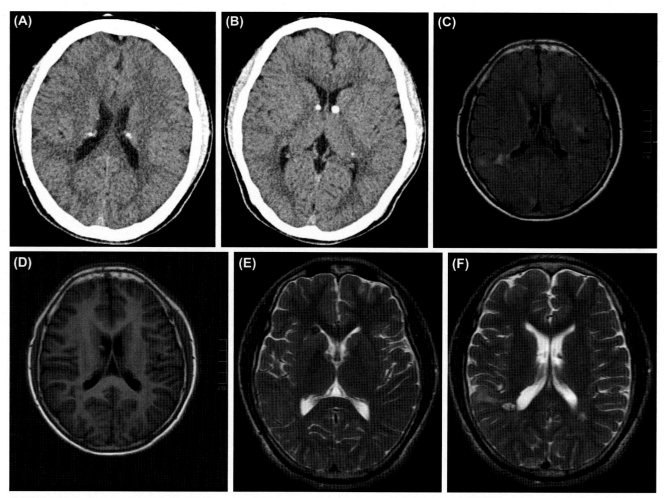

**FIGURE 13.2.2** Patient with tuberous sclerosis, female, 14 years old, with sudden disturbance of consciousness with convulsions for more than 1 day. (A—B). Head CT scan showing subependymal nodules. (C—F) axial FLAIR, T1WI, and T2WI image show subependymal nodules and abnormal signal in the voiceover mass of the lateral ventricle.

have obvious shagreen patches, which usually present as areas of thick leathery skin with a pebbly texture in the lumbosacral region. Ungual fibromas are small fleshy tumors occurring under fingernails or toenails and are present in about 20% of TSC patients. These dermatologic manifestations are considered major criteria in the diagnosis of TSC and are typically the only manifestations of TSC that can be detected at physical examination.

The characteristic pathological changes of the CNS of TSC are glial proliferative sclerosis tubers, which are mainly manifested as subependymal nodules, cortical and subcortical tubers, cerebral white matter radial migration lines, and subependymal giant cell astrocytomas. CT is more likely to identify calcified subependymal nodules, which are characterized by multiple nodular high-density shadows in the subependymal or ventricular margin, usually less than 5 mm in diameter (Fig. 13.2.2A—B). These lesions exhibit variable enhancement and appear hyperintense on T1-weighted MR images and low intensity on calcification, isointense to hyperintense on T2-weighted and FLAIR images when they have not yet calcified (Fig. 13.2.2C—F). Subependymal nodules can evolve into tumors, most of which occur in subependymal giant cell astrocytoma near the foramen of Monro with an incidence of 10%—15%.

# Further reading

[1] Henske Elizabeth P, Sergiusz J, Kingswood JC, et al. Tuberous sclerosis complex. Nat Rev Dis Prim 2016;2:16035.

[2] Paolo C, Roberta B, Sergiusz J. Tuberous sclerosis. Lancet 2008;372:657—68.

[3] Crino Peter B, Nathanson Katherine L, Henske EP. The tuberous sclerosis complex. N Engl J Med 2006;355:1345—56.

[4] Manoukian Saro B, Kowal Daniel J. Comprehensive imaging manifestations of tuberous sclerosis. AJR Am J Roentgenol 2015;204:933—43.

Chapter 13.3

# Sturge—Weber syndrome

Sturge—Weber syndrome (SWS) is a rare sporadic and congenital neurocutaneous syndrome defined by the association of a facial capillary malformation (port-wine stain [PWS]) in the ophthalmic distribution of the trigeminal nerve, with ipsilateral vascular glaucoma and vascular malformation of the eye, and vascular malformation of the brain (leptomeningeal angioma). The facial capillary malformation in SWS has often been referred to as facial PWS, due to the characteristic dark red color similar to the Portuguese liquor. The facial PWS typically involves the forehead and upper eyelid, in a distribution that resembles the area innervated by the first branch (ophthalmic branch) of the trigeminal nerve. "SWS" is the most largely used term; the syndrome is also known as encephalotrigeminal angiomatosis, encephalofacial angiomatosis, neuroophthalmo-cutaneous syndrome, extensive capillary-venous unilateral brain malformation, and so on.

PWS or nevus flammeus in SWS are well-demarcated red macular stains present at birth. With increasing age, the stain darkens in color and becomes raised and thickened. Sometimes multiple, small, red to purple nodules develop on the surface that confer a cobblestone pattern to the lesions. In a few instances, larger nodules develop representing pyogenic granulomas or acral arteriovenous tumors. Facial PWS usually have a sharp midline demarcation, although some extension over the midline may be observed (Fig. 13.3.1). PWS in patients with SWS always involve the skin innervated by the first arch of the trigeminal nerve (V1, forehead, and upper eyelid).

Seizures are the most common neurologic manifestations and have been reported to occur in 23%—83% of patients with SWS. Developmental delay and progressive mental retardation are reported in half of the patients with SWS. Recurrent headache is another common manifestation present in one third to one half of the patients. Radiologically, a leptomeningeal (pial) vascular malformation, commonly located in the parieto-occipital area, cerebral atrophy, and calcifications may be seen (Fig. 13.3.2). Cerebral lesions in SWS are ipsilateral to the facial PWS.

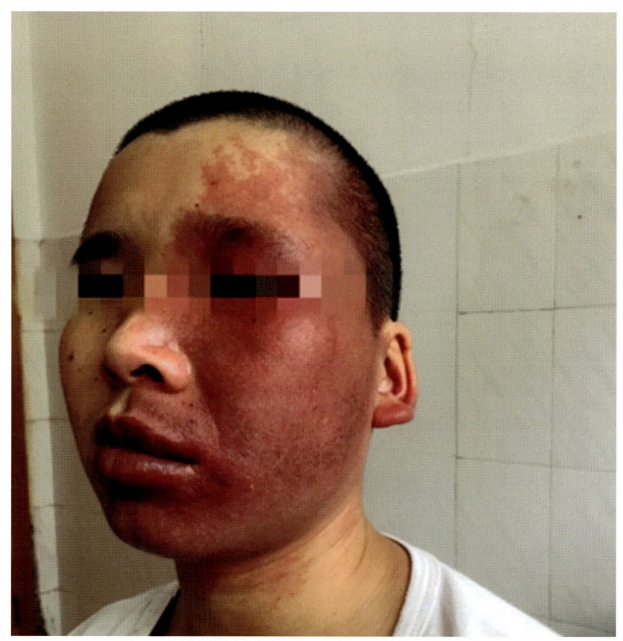

**FIGURE 13.3.1**  Patient with sturge–weber syndrome; port-wine stains are seen on the left side of the face.

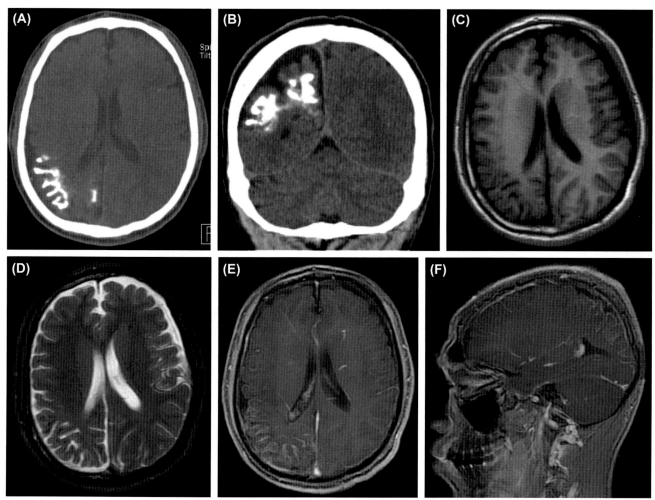

**FIGURE 13.3.2** (A−B) Plain CT scan shows cerebral atrophy in the right parietal occipital lobe with curvilinear, zonal calcifications along the gyri. (C−D) Axial T1WI and T2WI images show atrophy of the right parietal occipital lobe, widening of the sulcus, and calcification with hypointensity on T2WI. (E−F) Gadolinium-enhanced T1-weighted axial and sagittal images demonstrate asymmetric pia meningeal enhancement, which was linear and gyri-like meningeal enhancement.

## Further reading

[1] Baselga E. Sturge-Weber syndrome. Semin Cutan Med Surg June 2004;23(2):87−98.

[2] Bianchi F, Auricchio AM, Battaglia DI, et al. Sturge-Weber syndrome: an update on the relevant issues for neurosurgeons. Childs Nerv Syst October 2020;36(10):2553−70.

[3] Comi AM. Sturge-Weber syndrome. Handb Clin Neurol 2015;132:157−68.

[4] Hassanpour K, Nourinia R, Gerami E, et al. Ocular manifestations of the sturge-weber syndrome. J Ophthalmic Vis Res July 29, 2021;16(3):415−31.

[5] Sudarsanam A, Ardern-Holmes SL. Sturge-Weber syndrome: from the past to the present. Eur J Paediatr Neurol May 2014;18(3):257−66.

[6] Comi AM. Pathophysiology of sturge-weber syndrome. J Child Neurol August 2003;18(8):509−16.

Chapter 13.4

# Neurocutaneous melanosis

Neurocutaneous melanosis (NCM) is a rare congenital, nonfamilial sporadic syndrome characterized by the development of congenital melanocytic nevi (CMN) and benign or malignant melanocytic tumors of CNS. The disease usually occurs in Caucasians, and there is an equal gender predilection. NCM is characterized by three types of cutaneous lesions: (a) benign CMN; (b) proliferative melanocytic nodules; and eventually (c) malignant transformation into melanoma of previous lesions. Neurological manifestations are secondary to melanocytic proliferation, benign or malignant, of the CNS, including either the leptomeninges or brain parenchyma. The typical cutaneous lesions are recognized at birth; neurological manifestations usually appear later. There is no systemic involvement in NCM, or only rarely. Neurocutaneous melanosis may be a phakomatosis results from congenital dysplasia of the neuroectodermal melanocyte progenitor cells, leading to proliferation of melanin-producing cells in the skin and leptomeninges.

The skin lesions are present, and usually recognized, at birth, as single or multiple melanocytic nevi with abundant hair. Large CMN measure at least 20 cm in adults, extrapolated to at least 9 cm on the scalp or head in infants and 6 cm on the body. Approximately 80% of individuals with large or giant nevi present with associated smaller "satellite" nevi: the mean number of satellite nevi can range from 0 to 2,500, with a mean number of 80 and a median number of 20. The color of the skin complexion and hair is similar. In some neonates, this hyperpigmentation can fade over months resembling later to dark PWS: thus, a history of strong pigmentation at birth and concomitant hypertrichosis are of diagnostic importance. The nevus can cover 80% of the body surface: these patients are at higher risk for developing leptomeningeal and cutaneous melanoma. Isolated involvement of the extremities is the least frequent; in extensive nevi involving the extremities, atrophy (secondary to fat and/or bone reduced growth) of the affected limb can occur.

Neurologic symptoms result from symptomatic hydrocephalus that arises from proliferation of leptomeningeal deposits. Early neurologic symptoms typically consist of seizure and epilepsy. MRI reveals melanocytic deposits along leptomeninges and within parenchyma as hyperintensities on T1WI and hypointensities on T2WI. Enhancement with administration of contrast medium identifies early meningeal thickening. Enlarged subarachnoid spaces can also be evidence of meningeal thickening.

## Further reading

[1] Farabi B, Akay BN, Goldust M, et al. Congenital melanocytic naevi: an up-to-date overview. Australas J Dermatol May 2021;62(2):e178−91.
[2] Islam MP. Neurocutaneous melanosis. Handb Clin Neurol 2015;132:111−7.
[3] Chernoff KA, Schaffer JV. Cutaneous and ocular manifestations of neurocutaneous syndromes. Clin Dermatol 2016 ;34(2):183−204.
[4] Scattolin MA, Lin J, Peruchi MM, et al. Neurocutaneous melanosis: follow-up and literature review. J Neuroradiol December 2011;38(5):313−8.
[5] Jakchairoongruang K, Khakoo Y, Beckwith M, et al. New insights into neurocutaneous melanosis. Pediatr Radiol November 2018;48(12):1786−96.
[6] Ruggieri M, Polizzi A, Catanzaro S, et al. Neurocutaneous melanocytosis (melanosis). Childs Nerv Syst October 2020;36(10):2571−96.
[7] Smith AB, Rushing EJ, Smirniotopoulos JG. Pigmented lesions of the central nervous system: radiologic-pathologic correlation. Radiographics 2009 ;29(5):1503−24.
[8] DeDavid M, Orlow SJ, Provost N, et al. Neurocutaneous melanosis: clinical features of large congenital melanocytic nevi in patients with manifest central nervous system melanosis. J Am Acad Dermatol October 1996;35(4):529−38.

# Index

Printed in the United States
by Baker & Taylor Publisher Services